PERSPECTIVES ON SPIN GLASSES

Presenting and developing the theory of spin glasses as a prototype for complex systems, this book is a rigorous and up-to-date introduction to their properties.

The book combines a mathematical description with a physical insight of spin glass models. Topics covered include the physical origins of those models and their treatment with replica theory; mathematical properties such as correlation inequalities and their use in the thermodynamic limit theory; main exact solutions of the mean field models and their probabilistic structures; and the theory of the structural properties of the spin glass phase such as stochastic stability and the overlap identities. Finally, a detailed account is given of the recent numerical simulation results and properties, including overlap equivalence, ultrametricity, and decay of correlations. The book is ideal for mathematical physicists and probabilists working in disordered systems.

PIERLUIGI CONTUCCI is Professor of Mathematical Physics at the University of Bologna, and Research Director for the hard sciences section of the Istituto Cattaneo. His research interests are in statistical mechanics and its applications to socio-economic sciences.

CRISTIAN GIARDINÀ is Associate Professor in Mathematical Physics at the University of Modena and Reggio Emilia, and Visiting Professor in Probability at Nijmegen University. His research interests are in mathematical statistical physics and stochastic processes.

PERSPECTIVES ON SPIN GLASSES

PIERLUIGI CONTUCCI
University of Bologna

CRISTIAN GIARDINÀ
University of Modena and Reggio Emilia

Shaftesbury Road, Cambridge CB2 8EA, United Kingdom

One Liberty Plaza, 20th Floor, New York, NY 10006, USA

477 Williamstown Road, Port Melbourne, VIC 3207, Australia

314–321, 3rd Floor, Plot 3, Splendor Forum, Jasola District Centre, New Delhi – 110025, India

103 Penang Road, #05–06/07, Visioncrest Commercial, Singapore 238467

Cambridge University Press is part of Cambridge University Press & Assessment,
a department of the University of Cambridge.

We share the University's mission to contribute to society through the pursuit of
education, learning and research at the highest international levels of excellence.

www.cambridge.org
Information on this title: www.cambridge.org/9780521763349

First published 2013

A catalogue record for this publication is available from the British Library

Library of Congress Cataloging-in-Publication data
Contucci, Pierluigi, 1964–
Perspectives on spin glasses / Pierluigi Contucci and Cristian Giardinà.
pages cm
Includes bibliographical references and index.
ISBN 978-0-521-76334-9
1. Spin glasses – Mathematical models. I. Giardinà, Cristian. II. Title.
QC176.8.S68C667 2013
530.4′12 – dc23 2012032860

ISBN 978-0-521-76334-9 Hardback

Contents

Preface

Spin glasses are statistical mechanics systems with random interactions. The alternating sign of those interactions generates a complex physical behavior whose mathematical structure is still largely uncovered. The approach we follow in this book is that of mathematical physics, aiming at the rigorous derivation of their properties with the help of physical insight.

The book starts with the theoretical physics origins of the spin glass problem. The main models are introduced and a description of the replica approach is illustrated for the Sherrington–Kirkpatrick model.

Chapters 2 and 3 contain the starting points of the mathematical rigorous approach leading to the control of the thermodynamic limit for spin glass systems. Correlation inequalities are introduced and proved in various settings, including the Nishimori line. They are then used to prove the existence of the large-volume limit in both short-range and mean-field models.

Chapter 4 deals with exact results which belong to the mean-field case. The methods and techniques illustrated span from the Ruelle probability cascades to the Aizenman–Sims–Starr variational principle. In this framework the Guerra upper bound theorem for the pressure is presented and the Talagrand theorem is reported.

Chapter 5 deals with the structural identities characterizing the spin glass phase. These are obtained by an extension of the stochastic stability method, i.e. an invariance property of the system under small perturbations, together with the self-averaging property.

Chapter 6 features some problems which are still out of analytical reach and are investigated with numerical methods: the equivalence among different overlap structures, the hierarchical organization of the states, the decay of correlations, and the energy interface cost.

Needless to say there are innumerable important issues not covered by the book. Among them are the dynamical properties of spin glasses (see, for example, Sompolinsky and Zippelius (1982); Cugliandolo and Kurchan (1993); Bouchaud

(1992); Ben Arous *et al.* (2001, 2002); Bovier *et al.* (2001)). Another very large topic not covered is that of applications whose ideas originated within spin glass theory and successfully fertilized other areas.

It is a pleasure to thank the co-authors whose research work provided the foundations of this book: Michael Aizenman, Alessandra Bianchi, Mirko Degli Esposti, Claudio Giberti, Sandro Graffi, Andreas Knauf, Stefano Isola, Joel Lebowitz, Satoshi Morita, Hidetoshi Nishimori, Giorgio Parisi, Joe Pulé, Shannon Starr, Francesco Unguendoli, and Cecilia Vernia.

Useful conversations with many colleagues are acknowledged, particularly those with Louis-Pierre Arguin, Adriano Barra, Anton Bovier, Edouard Brézin, Aernout van Enter, Silvio Franz, Francesco Guerra, Frank den Hollander, Jorge Kurchan, Enzo Marinari, Marc Mezard, Chuck Newman, Dmitry Panchencko, Daniel Stein, and Michael Talagrand.

Last but not least we thank Claudio Giberti, Bernardo D'Auria, Aernout van Enter, and Cecilia Vernia for their careful reading of the manuscript.

1

Origins, models and motivations

Abstract

We introduce the basic spin glass models, namely the Edwards–Anderson model on a finite-dimensional lattice with short-range interaction and the Sherrington–Kirkpatrick model on the complete graph. The quenched equilibrium state which is used to describe the thermodynamical properties of a general disordered system is defined, together with the concept of real replicas. The notion of mean-field for a spin glass model is discussed. Finally, the original computations for the Sherrington–Kirkpatrick model based on the replica method are presented – namely the replica symmetric solution and the Parisi replica symmetry breaking scheme.

1.1 The spin glass problem

Spin glass models have been considered in different scientific contexts, including experimental condensed matter physics, theoretical physics, mathematical statistical physics and, more recently, probability. They have also been used to solve problems in fields as diverse as theoretical computer science (combinatorial optimization, traveling salesman, Boolean satisfiability, number partitioning, random assignment, error correcting codes, etc.), biology (Hopfield model), population genetics (hierarchical coalescence), and the economy (modelization of financial markets). Thus spin glasses represent a true example of a multi-disciplinary topic.

The study of spin glasses began after experiments on magnetic alloys, for instance metals like Fe, Mn and Cr weakly diluted in metals such as Au, Ag and Cu. It was observed that their thermodynamical behavior was not compatible with the theory of ferromagnetism and showed peculiar dynamical out-of-equilibrium properties such as aging and rejuvenation effects (for a recent account of spin glass dynamics and connection to experimental data see Cugliandolo and Kurchan (2008)). The experiments motivated the introduction of an Ising model with random interactions by Edwards and Anderson (1975). To simplify the model, and in the quest of a solvable model, a mean-field version was proposed by Sherrington and Kirkpatrick (1975). The mean-field theory was fully developed by G. Parisi, who proposed an ansatz to solve the problem exactly. Nowadays that theory is called *replica symmetry breaking* or *mean-field spin glass theory*. It revealed both unconventional physical properties and a very rich mathematical structure (Mézard *et al.* (1987)). In recent times some features of the theory have received a rigorous mathematical proof, in particular the computation of the free energy density in the thermodynamic limit due to Guerra (2003) and Talagrand (2006).

For the time being there is no consensus on the virtues of the mean-field Parisi solution in describing the behavior of magnetic alloys. While numerical simulations point to a mean-field behavior of the short-range Edwards–Anderson model on three-dimensional lattices, the mean-field picture has been questioned by the droplet-like picture in theoretical physics (Fisher and Huse (1988)) as well as by the metastate approach in mathematical physics (Newman and Stein (1996, 1998, 2002, 2003b); Newman (1997)).

Despite the lack of consensus about its relevance in condensed matter, in the last three decades the replica symmetry breaking theory has without doubt become a major paradigm in the theory of complex systems. It has been applied in the solution of many applied problems outside the realm of condensed matter physics, and the rich mathematical structure which has emerged from the Parisi solution of the Sherrington–Kirkpatrick (SK) model with non-rigorous techniques has attracted the interest of pure mathematicians and people working on rigorous results.

1.2 Random interactions, finite-dimensional models, mean-field models

The characteristic property of spin glass models is the presence of both positive ferromagnetic and negative antiferromagnetic interactions between the spins. While ferromagnetic couplings force the alignment of the spins in the low temperature phase, antiferromagnetic couplings prefer to anti-align the spins. When all bonds between spins cannot be satisfied, the model is generically said to be frustrated. For a precise definition of the concept of frustration one can look at the original paper by Toulouse (1977); the simplest example of a model containing apparent disorder – which can actually be removed by a gauge transform – is given in Mattis (1976).

Frustration can be realized in *deterministic* systems by properly choosing the couplings. An interesting class of deterministic systems with frustration and consequent glassy behavior is given by the so-called "sine model", introduced in

Bouchaud and Mézard (1994); Marinari *et al.* (1994a, b) and further studied in Degli Esposti *et al.* (2001, 2003); Contucci *et al.* (2002).

However, the standard spin glass models have *random* interactions. Three distinct classes of models are usually considered:

- *finite-dimensional spin glasses* are defined on a d-dimensional lattice (with d an integer number) and typically have finite-range interactions among the spins;
- *mean-field spin glasses* are defined on the complete graph with interactions between all pairs (or k-tuples, for an integer number k) of spins;
- *spin glasses on random graphs* are defined on a random graph with interaction between the spins linked by an edge.

Random graphs constitute a very interesting subject per se. The simplest example is the Erdős–Rényi random graph with edges which are independently present with identical probability. More general random graphs, such as the configuration model or the preferential attachment model, also include dependence structures showing power law degree distribution and small-world effects (see for instance the lecture notes by van der Hofstad (2012)). Spin glasses on random graphs therefore have a double source of randomness, given by the spatial structure where the interaction takes place, and the sign and magnitude of the couplings between spins. They will not be investigated in this book; the interested reader may consult Mézard and Montanari (2009).

In this book we shall focus on the first two classes of spin glass models, showing whenever possible their differences and similarities. We now define the primary examples of each class.

Definition 1.1 (Edwards–Anderson Model) Consider a system in a box $\Lambda \subset \mathbb{Z}^d$ made of interacting spins $\sigma = \{\sigma_i\}_{i\in\Lambda}$ with $\sigma_i \in \{-1, +1\}$; the Edwards–Anderson model is defined by the Hamiltonian

$$H_\Lambda(\sigma, J) = -\sum_{||i-j||=1} J_{i,j}\sigma_i\sigma_j, \tag{1.1}$$

where $||\cdot||$ denotes Euclidean distance and the couplings $J = \{J_{i,j}\}$ are independent random variables, all having the same distribution, which are assumed to be symmetric with

$$\mathbb{E}\left[J_{i,j}\right] = 0 \qquad \mathbb{E}\left[J_{i,j}^2\right] = 1 \tag{1.2}$$

where $\mathbb{E}\left[\cdot\right]$ denotes expectation. Note that the sum in the Hamiltonian is restricted to pairs of nearest-neighbor sites. A straightforward computation based on (1.2) and on independence gives, for the covariance of the Hamiltonian (a family of $2^{|\Lambda|}$

centered random variables),

$$\mathbb{E}(H_\Lambda(\sigma, J)H_\Lambda(\tau, J)) = d|\Lambda|Q_\Lambda(\sigma, \tau), \tag{1.3}$$

where $Q_\Lambda(\sigma, \tau)$ is the *bond* overlap between two spin configurations σ and τ and is given by

$$Q_\Lambda(\sigma, \tau) = \frac{1}{d|\Lambda|} \sum_{||i-j||=1} \sigma_i \sigma_j \tau_i \tau_j. \tag{1.4}$$

In the case of standard Gaussian distributed interactions $\{J_{i,j}\}$, the previous formula for the covariance completely identifies the model and can be used as an alternative definition.

Remark 1.2 Boundary conditions do matter both in (1.1) and (1.4). This will be analyzed in Chapter 3.

Definition 1.3 (SK model) Consider a system of N spins $\sigma = \{\sigma_i\}_{i\in\{1,\dots,N\}}$ with $\sigma_i \in \{-1, +1\}$; the SK model is defined by the Hamiltonian

$$H_N(\sigma, J) = -\frac{1}{\sqrt{2N}} \sum_{i,j=1}^{N} J_{i,j}\sigma_i\sigma_j, \tag{1.5}$$

where, for each couple $(i, j) \in \{1, \dots, N\}^2$, the couplings $J = \{J_{i,j}\}$ are a family of independent identical random variables with symmetric distribution, and $\mathbb{E}\left[J_{i,j}\right] = 0$ and $\mathbb{E}\left[J_{i,j}^2\right] = 1$.

If the couplings $\{J_{i,j}\}$ have a standard Gaussian distribution, then an equivalent definition is that the energy levels of the SK model in the volume $\{1, \dots, N\}$ are given by a family of 2^N centered Gaussian random variables with covariance

$$\mathbb{E}(H_N(\sigma, J)H_N(\tau, J)) = \frac{N}{2} q_N^2(\sigma, \tau), \tag{1.6}$$

where $q_N(\sigma, \tau)$ is the *site* overlap between two spin configurations σ and τ and is given by

$$q_N(\sigma, \tau) = \frac{1}{N} \sum_{i=1}^{N} \sigma_i \tau_i. \tag{1.7}$$

In this book we will often work with a general spin glass model which includes (as special cases) both the Edwards–Anderson model and the SK model, as well as other finite-dimensional or mean-field models which will be defined later. If the couplings are chosen to have a Gaussian distribution then it is possible to use properties of Gaussian random variables (such as the integration by parts formula) which allow a few simplifications in the computations. The construction of such a model will be discussed in Chapter 2 where a general representation theorem for Gaussian Hamiltonians will be presented. The condition on the model parameters (means and variances of the couplings) for the thermodynamic limit to exist will be analyzed in Chapter 3. Here we limit ourselves to the following definition.

Definition 1.4 (General spin glass model) For a volume $\Lambda \subset \mathbb{Z}^d$ and spins $\sigma_i = \{-1, +1\}$ sitting on every site $i \in \Lambda$, the general spin glass model is defined by the Hamiltonian

$$H_\Lambda(\sigma, J) = -\sum_{X \subset \Lambda} J_X \sigma_X, \tag{1.8}$$

where $\sigma_X = \prod_{i \in X} \sigma_i$ and the couplings $J = \{J_X\}_{X \in \Lambda}$ are independent random variables (with $J_\emptyset = 0$).

If those random variables are chosen to have a centered Gaussian distribution with variance $\mathbb{E}(J_X^2) = \Delta_X^2$, then an equivalent definition of the model is given by a family of $2^{|\Lambda|}$ centered Gaussian random variables $H_\Lambda(\sigma, J)$ with covariance

$$\mathbb{E}(H_\Lambda(\sigma, J) H_\Lambda(\tau, J)) = \mathcal{C}(\sigma, \tau) = |\Lambda| c_\Lambda(\sigma, \tau), \tag{1.9}$$

where $c_\Lambda(\sigma, \tau)$ is the *generalized* overlap between the two spin configurations σ and τ and is given by

$$c_\Lambda(\sigma, \tau) = \frac{1}{|\Lambda|} \sum_{X \subset \Lambda} \Delta_X^2 \sigma_X \tau_X. \tag{1.10}$$

Remark 1.5 By Schwartz' inequality, $|c_\Lambda(\sigma, \tau)| \leqslant c_\Lambda(\sigma, \sigma)$. A sufficient condition to guarantee existence of the thermodynamic limit is $\sup_\Lambda c_\Lambda(\sigma, \sigma) \leqslant \bar{c} < +\infty$ (see Section 3.2). Without loss of generality we will often assume that $c_\Lambda(\sigma, \sigma) = 1$.

Remark 1.6 The Edwards–Anderson model and the SK model correspond to special choices of the volume Λ and of the centered couplings J_X in the general spin glass model. Namely:

1. Definition 1.1 is recovered with the choice: $\Lambda \subset \mathbb{Z}^d$, and $\Delta_X = 1$ for $X = (i, j)$ with $(i, j) \in \mathbb{Z}^d \times \mathbb{Z}^d$ and $||i - j|| = 1$, and $\Delta_X = 0$ otherwise.
2. Definition 1.3 is recovered with the choice: $\Lambda = \{1, \ldots, N\}$, and $\Delta_X = \frac{1}{\sqrt{2N}}$ for $X = (i, j)$ with $(i, j) \in \{1, \ldots, N\}^2$, and $\Delta_X = 0$ otherwise.

Note however that the differences between finite-range interactions (constant variances Δ_X^2) and infinite-range interactions (variances Δ_X^2 depend on the volume) could imply substantial differences a priori on the thermodynamic properties in the large-volume limit.

1.3 Quenched measure and real replicas

The description of the thermodynamic properties of a disordered system with a random Hamiltonian requires the introduction of the *quenched state* notion. In spin glasses, the timescale of the spin variables' relaxation was observed to be much shorter than that of the interaction variables. This dynamical feature led physics to consider the interaction coefficient as *frozen* with respect to the spin ones. A proper mathematical formulation is then obtained by first averaging over the spin variables and computing the Boltzmann–Gibbs expectations, and then averaging over the disorder.

Definition 1.7 (Quenched expectation) For a random Hamiltonian of the form (1.8) on the volume Λ and a random (i.e. possibly depending on the J) function $f : \Sigma_\Lambda \to \mathbb{R}$ with $\Sigma_\Lambda = \{-1, +1\}^{|\Lambda|}$, the expectation with respect to the *random* Boltzmann–Gibbs measure at inverse temperature $\beta \geqslant 0$ is given by

$$\omega_{\Lambda,\beta}(f) = \frac{\sum_{\sigma\in\Sigma_\Lambda} f(\sigma)\exp[-\beta H_\Lambda(\sigma, J)]}{\sum_{\sigma\in\Sigma_\Lambda} \exp[-\beta H_\Lambda(\sigma, J)]}. \tag{1.11}$$

Averaging over the disorder, one obtains the *quenched* expectation, denoted by

$$\langle f\rangle_{\Lambda,\beta} = \mathbb{E}\left[\omega_{\Lambda,\beta}(f)\right]. \tag{1.12}$$

Remark 1.8 To alleviate the notation, we will not always write explicitly the volume- or temperature-dependence of either the random Boltzmann–Gibbs expectations or the quenched expectations.

Moreover, it is useful to distinguish between random thermodynamic observables and their quenched average. The fundamental thermodynamical quantity is the pressure (which gives up to a factor $1/\beta$ the negative of the free energy).

Definition 1.9 (Pressure) We define the *random partition function*

$$\mathcal{Z}_\Lambda(\beta) = \sum_{\sigma\in\Sigma_\Lambda} \exp[-\beta H_\Lambda(\sigma, J)], \tag{1.13}$$

the *random pressure*

$$\mathcal{P}_\Lambda(\beta) = \log \mathcal{Z}_\Lambda(\beta) \tag{1.14}$$

and the *quenched pressure*

$$P_\Lambda(\beta) = \mathbb{E}[\mathcal{P}_\Lambda(\beta)]. \tag{1.15}$$

Remark 1.10 In the above definition we assume *free boundary conditions.* We will return to the choice of boundary conditions in Chapter 3, where we will show that they do not matter for the thermodynamic limit of the pressure and we will analyze the effect of them on the surface pressure in Section 3.8.

The generalized overlap in Eq. (1.10) is the main observable of spin glass theory. Indeed the standard thermodynamic quantities can be expressed in terms of the quenched expectation of the generalized overlaps among copies of the system, called *real replicas,* all subject to the same disorder. It is thus useful to introduce the product random Boltzmann–Gibbs state over real replicas and the corresponding quenched state.

Definition 1.11 (Real replicas) For a random Hamiltonian of the form (1.8) on the volume Λ and a random function $f : \Sigma_\Lambda^R \to \mathbb{R}$, the expectation with respect to the R-product random Boltzmann–Gibbs state (with R being an integer) is

$$\Omega_{\Lambda,\beta}(f) = \sum_{\{\sigma^{(1)},\ldots,\sigma^{(R)}\}\in\Sigma_\Lambda^R} \frac{f(\sigma^{(1)},\ldots,\sigma^{(R)})e^{-\beta[H_\Lambda(\sigma^{(1)},J)+\cdots+H_\Lambda(\sigma^{(R)},J)]}}{[\mathcal{Z}_\Lambda(\beta)]^R}, \quad (1.16)$$

The *quenched* expectation is then

$$\langle f\rangle_{\Lambda,\beta} = \mathbb{E}\left[\Omega_{\Lambda,\beta}(f)\right]. \quad (1.17)$$

We now show how the concept of real replicas, which might seem artificial at first sight, naturally arises in the expression of the main thermodynamic quantities in the context of Gaussian spin glass models. Consider, for instance, the *internal energy* given by:

$$U_\Lambda(\beta) = -\frac{dP_\Lambda}{d\beta}(\beta) = \mathbb{E}(\omega_{\Lambda,\beta}(H_\Lambda)). \quad (1.18)$$

Using the integration by parts formula for a set of centered Gaussian random variables $X = (X_1, \ldots, X_k)$ with covariances $a_{i,j} = \mathbb{E}\left[X_i X_j\right]$, namely

$$\mathbb{E}(X_i f(X)) = \sum_{j=1}^{n} a_{i,j}\,\mathbb{E}\left(\frac{\partial f}{\partial x_j}(X)\right), \quad (1.19)$$

and assuming $c_\Lambda(\sigma,\sigma) = 1$, one obtains the following expression for the internal energy:

$$U_\Lambda(\beta) = -\beta|\Lambda|\langle 1 - c_{1,2}\rangle_{\Lambda,\beta}, \quad (1.20)$$

with

$$\langle c_{1,2}\rangle_{\Lambda,\beta} = \mathbb{E}\left[\sum_{\{\sigma^{(1)},\sigma^{(2)}\}\in\Sigma_\Lambda^2} c_\Lambda(\sigma^{(1)},\sigma^{(2)})\frac{e^{-\beta[H_\Lambda(\sigma^{(1)},J)+H_\Lambda(\sigma^{(2)},J)]}}{[\mathcal{Z}_\Lambda(\beta)]^2}\right]. \quad (1.21)$$

Another example is the *specific heat*

$$C_\Lambda(\beta) = -\beta^2 \frac{dU_\Lambda}{d\beta}(\beta) = \beta^2\, \mathbb{E}\big[\omega_{\Lambda,\beta}\big(H_\Lambda^2\big) - \omega_{\Lambda,\beta}^2(H_\Lambda)\big]. \tag{1.22}$$

As for the internal energy, now using integration by parts twice, one obtains the following result:

$$C_\Lambda(\beta) = \beta^2|\Lambda|\langle 1 - c_{1,2}\rangle_{\Lambda,\beta} - 2\beta^4|\Lambda|^2\big\langle c_{1,2}^2 - 4c_{1,2}c_{2,3} + 3c_{1,2}c_{3,4}\big\rangle_{\Lambda,\beta}, \tag{1.23}$$

with

$$\langle c_{1,2}c_{2,3}\rangle_{\Lambda,\beta} = \mathbb{E}\Bigg[\sum_{\{\sigma^{(1)},\sigma^{(2)},\sigma^{(3)}\}\in\Sigma_\Lambda^3} c_\Lambda(\sigma^{(1)},\sigma^{(2)})c_\Lambda(\sigma^{(2)},\sigma^{(3)}) \times \frac{e^{-\beta[H_\Lambda(\sigma^{(1)},J)+H_\Lambda(\sigma^{(2)},J)+H_\Lambda(\sigma^{(3)},J)]}}{[\mathcal{Z}_\Lambda(\beta)]^3}\Bigg], \tag{1.24}$$

and an analogous expression (involving four replicas) for $\langle c_{1,2}c_{3,4}\rangle_{\Lambda,\beta}$.

In the previous formula we used the same bracket symbol as (1.17) in the lefthand sides of (1.21) and (1.24) where the observable f was a function of the spin configurations. The precise meaning of such a small abuse of notation is given by the following:

Definition 1.12 (Generalized overlaps random variables) For any integer $R \geqslant 1$ the formulas (1.21) and (1.24) and their generalization to arbitrary powers and an arbitrary number R of copies, define the family of random variables $\{c_{l,m}\}$ with $1 \leqslant l < m \leqslant R$ and their joint distribution $p_{lm}^{(\Lambda)}, p_{lm,l'm'}^{(\Lambda)}, \ldots$ via

$$\big\langle c_{1,2}^k\big\rangle_\Lambda = \int dx x^k p_{12}^{(\Lambda)}(x) \tag{1.25}$$

$$\big\langle c_{1,2}^k c_{2,3}^l\big\rangle_\Lambda = \int dx \int dy x^k y^l p_{12,23}^{(\Lambda)}(x, y) \tag{1.26}$$

and similar. What the spin glass theory is interested in is the behavior of the former random variables in the thermodynamic limit whose distribution we will denote by $p_{12}(x)$, $p_{12,23}(x, y)$, etc.

Remark 1.13 We point out that the joint distribution of the overlap random variables is invariant under the action of the permutation group on the set $\{1, \ldots, \mathrm{R}\}$. Denoting the symmetric random matrix with elements $\{c_{l,m}\}$ by C and assuming without loss of generality $c_{l,l} = 1$, the invariance under the permutation group is expressed by

$$PCP^{-1} \stackrel{\mathcal{D}}{=} C \qquad \text{for all } P, \tag{1.27}$$

where P is the matrix associated to a permutation of the set $\{1, 2, \ldots, R\}$. For example:

$$p_{12}(x) = p_{56}(x), \qquad p_{12,23}(x, y) = p_{13,35}(x, y). \tag{1.28}$$

1.4 Definition of a mean-field spin glass

A spin glass model, such as the SK model, was given the name *mean-field* type for reasons of similarity with the Curie–Weiss model of ferromagnetism, in which the spins' interaction space is the complete graph of N vertices and the mutual interaction is invariant with respect to the permutation group. A similar property is also true for the SK model since the interaction space is still the complete graph and the interactions are invariant, in distribution, under the action of the permutation group. The apparent similarity between the two models goes much further and is manifest in the replica symmetric solution of the SK model, to the point that one could think of this model as a random version of the Curie–Weiss model. The similarity goes as follows.

The magnetization of the Curie–Weiss model, defined as

$$m = \lim_{N\to\infty} \omega_N^{(CW)} \left(\frac{1}{N} \sum_{i=1}^{N} \sigma_i \right) \tag{1.29}$$

where $\omega_N^{CW}(\cdot)$ denotes expectation with respect to the Boltzmann–Gibbs measure with Hamiltonian

$$H_N^{(CW)}(\sigma) = -\frac{1}{2N} \sum_{i,j=1}^{N} \sigma_i \sigma_j - h \sum_{i=1}^{N} \sigma_i, \tag{1.30}$$

satisfies the equation (at positive inverse temperature β)

$$m = \tanh(\beta[h + m]). \tag{1.31}$$

This equation can also be obtained from the mean-field ferromagnetic model with Hamiltonian

$$H_N^{(MF)}(\sigma) = -\sum_{i=1}^{N} \sigma_i (h + M) \tag{1.32}$$

where M is a parameter representing the average field which is caused by all the other spins and which is required to satisfy the self-consistency equation

$$M = \omega_N^{(MF)}\left(\frac{1}{N}\sum_{i=1}^{N}\sigma_i\right), \tag{1.33}$$

with $\omega_N^{(MF)}(\cdot)$ the expectation with respect to the Boltzmann–Gibbs measure with Hamiltonian (1.32). An immediate computation shows that the self-consistency equation (1.33) is equivalent to Eq. (1.31) with $M = m = \lim_{N\to\infty}\omega_N^{(CW)}(\sigma_i)$. Therefore, one usually says that the Curie–Weiss model is the mean-field theory of ferromagnetism.

Considering now the SK model, we will see in Section 1.6 that the original replica symmetric solution yields the following equation for the quenched expectation of the overlap:

$$q = \int d\mu(z)\tanh^2(\beta(h + \sqrt{q}z)), \tag{1.34}$$

where $d\mu(z) = \frac{1}{\sqrt{2\pi}}e^{-\frac{z^2}{2}}dz$ and

$$q = \lim_{N\to\infty}\left\langle\frac{1}{N}\sum_{i=1}^{N}\sigma_i\tau_i\right\rangle_N^{(SK)} = \lim_{N\to\infty}\mathbb{E}\left(\frac{1}{N}\sum_{i=1}^{N}\left[\omega_N^{(SK)}(\sigma_i)\right]^2\right), \tag{1.35}$$

with $\langle\cdot\rangle_N^{(SK)}$ denoting the quenched expectation associated to the Hamiltonian

$$H_N^{(SK)}(\sigma) = -\frac{1}{\sqrt{2N}}\sum_{i,j=1}^{N}J_{i,j}\sigma_i\sigma_j - h\sum_{i=1}^{N}\sigma_i \tag{1.36}$$

and $\{J_{i,j}\}$ independent and identically distributed (i.i.d.) standard Gaussian random variables. Equation (1.34) can also be obtained from non-interacting spin models of the form

$$\tilde{H}_N^{(MF)}(\sigma) = -\sum_{i=1}^{N}\sigma_i(h + M_i) \tag{1.37}$$

where each spin σ_i, besides the external field h, feels the action of a centered Gaussian random field M_i with covariance

$$\mathbb{E}(M_iM_j) = \delta_{i,j}\bar{M}^2 \tag{1.38}$$

where $\bar{M}$ is determined self-consistently by imposing that

$$\bar{M}^2 = \left\langle\frac{1}{N}\sum_{i=1}^{N}\sigma_i\tau_i\right\rangle_N^{(MF)} \tag{1.39}$$

where $\langle\cdot\rangle_N^{(MF)}$ is the quenched state relative to the Hamiltonian (1.37). An immediate computation shows that the self-consistency equation (1.39) is equivalent to (1.34) with $\bar{M}^2 = q = \lim_{N\to\infty} \mathbb{E}([\omega_N^{(SK)}(\sigma_i)]^2)$.

With the parallelism described above, one could say that the replica symmetric solution of the SK model is the *mean-field* theory of spin glasses, described by a single number (the average overlap q) as the Curie–Weiss model is the mean-field theory of ferromagnets for which all information is encoded in the magnetization m. However, Eq. (1.34) for the quenched expectation of the overlap in the SK model is not correct below the critical temperature. As stated in the papers by Thouless *et al.* (1977) and Almeida and Thouless (1978), immediately after the replica symmetric solution given in Sherrington and Kirkpatrick (1975), the effective field acting on a single spin of the SK model is far from having a Gaussian distribution. Indeed, as will be shown in Section 1.6, the assumption of a Gaussian distribution for the effective field leads to unphysical properties like negative entropy.

A more effective notion of mean-field for the Curie–Weiss model can be stated as follows: in the presence of an external field $h \neq 0$, one has the property:

$$\lim_{N\to\infty}\left[\omega_N^{(CW)}\left(\left(\frac{1}{N}\sum_{i=1}^N \sigma_i\right)^2\right) - \left(\omega_N^{(CW)}\left(\frac{1}{N}\sum_{i=1}^N \sigma_i\right)\right)^2\right] = 0. \tag{1.40}$$

Using translation invariance, this amounts to the independence property of the spins in the thermodynamic limit with respect to the Boltzmann–Gibbs measure, namely (with $h \neq 0$ and m the solution of (1.31))

$$\lim_{N\to\infty} \omega_N^{(CW)}(\sigma_1\sigma_2) = \lim_{N\to\infty}\left(\omega_N^{(CW)}(\sigma_1)\right)^2 = m^2 \tag{1.41}$$

or more generally, for $1 \leqslant i_1 < i_2 < \cdots < i_k \leqslant N$,

$$\lim_{N\to\infty} \omega_N^{(CW)}(\sigma_{i_1}\sigma_{i_2}\cdots\sigma_{i_k}) = \lim_{N\to\infty}\left(\omega_N^{(CW)}(\sigma_1)\right)^k = m^k. \tag{1.42}$$

From the equations above it is evident that the order parameter of the mean-field ferromagnetic theory is the magnetization function of the temperature.

Similarly, for the SK model at high temperatures, a simple factorization rule holds due to the self-averaging property of the overlap between two copies:

$$\lim_{N\to\infty}\left[\left\langle\left(\frac{1}{N}\sum_{i=1}^N \sigma_i\tau_i\right)^2\right\rangle_N - \left\langle\frac{1}{N}\sum_{i=1}^N \sigma_i\tau_i\right\rangle_N^2\right] = 0, \tag{1.43}$$

which implies that, for $h \neq 0$ and q the solution of (1.34),

$$\lim_{N\to\infty}\langle\sigma_1\tau_1\sigma_2\tau_2\rangle_N = \lim_{N\to\infty}\langle\sigma_1\tau_1\rangle_N^2 = q^2 \tag{1.44}$$

or more generally, for $1 \leqslant i_1 < i_2 < \cdots < i_k \leqslant N$,

$$\lim_{N\to\infty} \langle \sigma_{i_1}\tau_{i_1}\sigma_{i_2}\tau_{i_2}\cdots\sigma_{i_k}\tau_{i_k}\rangle_N = \lim_{N\to\infty} \langle \sigma_1\tau_1\rangle_N^k = q^k. \tag{1.45}$$

These are indeed the factorization rules assumed in the replica symmetric solution which fails at low temperatures.

However, the Parisi solution of the SK model is precisely based on the idea that the overlap is *not* a self-averaging quantity, i.e. it fluctuates with respect to the quenched measure even in the thermodynamic limit[1].

A natural question then follows: is it possible to identify a factorization rule which holds for the SK model at all temperatures? Such a rule must coincide with Eq. (1.45) at high temperatures but not at low temperatures. Since the relevant observable of the theory is the overlap, one could ask what is the factorization rule for the joint distribution of the overlaps among many copies? The remarkable fact that was found in the Parisi solution is that *all joint overlap distributions can be expressed in terms of the distribution of the single overlap*. Denoting by $p(x)$ the overlap distribution of any two copies (see Remark 1.13), the simplest examples involving two overlaps are the following:

$$p_{12,23}(x,y) = \frac{1}{2}p(x)\delta(x-y) + \frac{1}{2}p(x)p(y) \tag{1.46}$$

$$p_{12,34}(x,y) = \frac{1}{3}p(x)\delta(x-y) + \frac{2}{3}p(x)p(y). \tag{1.47}$$

Probably the most celebrated factorization rule is the *ultrametric* relation for the joint distribution of the overlaps of three real replicas:

$$\begin{aligned} p_{12,23,31}(x,y,z) = {} & \frac{1}{2}p(x)\chi(x)\delta(x-y)\delta(y-z) \\ & + \frac{1}{2}p(x)p(y)\theta(x-y)\delta(y-z) \\ & + \frac{1}{2}p(y)p(z)\theta(y-z)\delta(z-x) \\ & + \frac{1}{2}p(z)p(x)\theta(z-x)\delta(x-y), \end{aligned} \tag{1.48}$$

where $\chi(x) = \int_0^x p(x')dx'$ and $\theta(x)$ denote the Heaviside step function. The previous examples are only the lowest cases of the general factorization rule structure for the overlaps of an arbitrary number of real replicas discovered in the replica symmetry breaking solution or mean-field solution of the SK model (Mézard *et al.*

[1] It is interesting to remark, however, that the overlap can be non-self-averaging even without disorder, as shown in non-periodic sequences by van Enter *et al.* (1992); van Enter and de Groote (2011).

(1987)). They allow the reconstruction of the entire joint distribution of the overlaps, starting from the single overlap distribution.

Definition 1.14 (Mean-field spin glass) A spin glass model is of *mean-field* type if the following two properties hold:

- the single overlap probability distribution is nontrivial at low temperatures $T \leqslant T_c$;
- the joint probability distribution of the overlaps satisfies the factorization rules of the Parisi solution of the SK model.

1.5 Replica method for the SK model

Much of the interest around spin glasses (Aizenman *et al.* (1987)) has been motivated by the Parisi solution of the SK model. We devote the last part of this chapter to the presentation of those results.

In the mean-field theory of ferromagnets, the order parameter is a scalar quantity (e.g. the spontaneous magnetization) which is zero above the critical temperature and different from zero in the low temperature phase. The free energy is expressed as a variational problem involving the minimization of a real function. The minimum, which satisfies the self-consistent equation (1.31), gives the equilibrium value of the magnetization.

In the Parisi solution of the SK model, the order parameter is a *function*. This function is actually the cumulative distribution function of a random variable which is the overlap between two real replicas. In the physics literature, the order parameter of the SK model is called a "functional order parameter" to stress the fact that the free energy is expressed as a variational problem involving the extremality of a functional. Moreover in the Parisi formulation the free energy must be *maximized* rather than minimized. The aim of this and the following sections is to show the original ingenious computations which lead to the correct expression of the free energy of the SK model using the replica method (Parisi (1979a, b, 1980a, b, c, 1983)).

We recall that the Hamiltonian of the SK model is given in Definition 1.3. It is convenient to also include in the Boltzmann factor a term like $h\sum_{i=1}^{N}\sigma_i$ proportional to an external field h. The finite-volume quenched pressure is given by

$$p_N(\beta, h) = \frac{1}{N}\mathbb{E}\left[\ln \mathcal{Z}_N(\beta, h, J)\right]. \tag{1.49}$$

In order to compute the infinite-volume quenched pressure, physicists have introduced the *replica method*, which works as follows. Define, for a given volume N,

$$p_N(\beta, h, n) = \frac{1}{Nn} \ln \mathbb{E}\left([\mathcal{Z}_N(\beta, h, J)]^n\right). \tag{1.50}$$

For positive integers n, this n-dependent pressure is much easier to calculate than the quenched pressure, because the average is now inside the logarithm. To recover the finite-volume quenched pressure, one can extend the definition (1.50) to real values of n and use the following identity:

$$p_N(\beta, h) = \lim_{n\to 0} p_N(\beta, h, n), \tag{1.51}$$

which can be proved using the de l'Hôpital theorem. Moreover, by a simple interpolation technique that will be described later in Chapter 3, one can establish the existence of the thermodynamic limit of the n-dependent pressure

$$p(\beta, h, n) = \lim_{N\to\infty} p_N(\beta, h, n). \tag{1.52}$$

To compute the thermodynamic limit of the quenched pressure one might proceed by assuming that the two limits $N \to \infty$ and $n \to 0$ can be interchanged, thus obtaining

$$\begin{aligned} p(\beta, h) &= \lim_{N\to\infty} p_N(\beta, h) \\ &= \lim_{N\to\infty} \lim_{n\to 0} p_N(\beta, h, n) \\ &= \lim_{n\to 0} \lim_{N\to\infty} p_N(\beta, h, n) \\ &= \lim_{n\to 0} p(\beta, h, n). \end{aligned} \tag{1.53}$$

Therefore, the replica method works as follows: first compute $p(\beta, h, n) = \lim_{N\to\infty} p_N(\beta, h, n)$ for integer n; then make an analytical continuation to real n; finally take the limit $n \to 0$.

In the case of the SK model, one has the following results using the replica method.

Replica computation 1 (n-dependent pressure) *For the SK model with Hamiltonian (1.5), for integer $n \geqslant 1$, one has*

$$p(\beta, h, n) = \sup_Q (A(Q, \beta, h, n)) \tag{1.54}$$

with

$$A(Q, \beta, h, n) = -\frac{\beta^2}{4n} \sum_{a,b=1}^{n} Q_{a,b}^2 + \frac{1}{n} \ln \left(\sum_{\{S_1,\dots,S_n\}\in\Sigma_n} e^{\beta h \sum_{a=1}^{n} S_a + \frac{\beta^2}{2} \sum_{a,b=1}^{n} S_a S_b Q_{a,b}} \right) \tag{1.55}$$

and Q an $n \times n$ symmetric matrix with real elements $Q_{a,b}$, and on the diagonal $Q_{a,a} = 1$.

Proof To obtain (1.54) one begins by observing that the integer moments of the finite-volume partition function are given by

$$[\mathcal{Z}_N(\beta, h, J)]^n = \sum_{\{\sigma^{(1)},\dots,\sigma^{(n)}\}\in\Sigma_N^n} e^{\beta h \sum_{i=1}^N \sum_{a=1}^n \sigma_i^{(a)}} e^{-\beta(H_N(\sigma^{(1)},J)+\cdots+H_N(\sigma^{(n)},J))},$$

where the first sum is over the spins of n replicas, each of volume N. The disorder average can be computed using the formula for the moment-generating function of a centered k-dimensional Gaussian vector $X = (X_1, \dots, X_k)$ with covariances $a_{i,j} = \mathbb{E}\left[X_i X_j\right]$

$$\mathbb{E}\left(e^{\sum_{i=1}^k t_i X_i}\right) = e^{\frac{1}{2}\sum_{i,j=1}^k a_{i,j} t_i t_j}. \tag{1.56}$$

Using (1.6) one obtains

$$\mathbb{E}([\mathcal{Z}_N(\beta, h, J)]^n) = \sum_{\{\sigma^{(1)},\dots,\sigma^{(n)}\}\in\Sigma_N^n} e^{\beta h \sum_{i=1}^N \sum_{a=1}^n \sigma_i^{(a)}} e^{\frac{N\beta^2}{4}\sum_{a,b=1}^n [q_N(\sigma^{(a)},\sigma^{(b)})]^2}. \tag{1.57}$$

At this point the problem is deterministic. One could fix all the overlaps $q_N(\sigma^{(a)}, \sigma^{(b)})$ and compute the number of spin configurations corresponding to those fixed values, and then apply the saddle-point method to compute the result in the thermodynamic limit. However, it is more convenient to proceed via the Hubbard–Stratonovich formula (i.e. turning around the formula (1.56) for the moment-generating function); so we introduce new variables $Q_{a,b}$ that allow the quadratic overlaps to linearize in the exponent. Namely, for $a \neq b$

$$e^{\frac{N\beta^2}{4}[q_N(\sigma^{(a)},\sigma^{(b)})]^2} = \int_{-\infty}^{+\infty} dQ_{a,b}\sqrt{\frac{\beta^2 N}{4\pi}}\, e^{-\frac{\beta^2 N}{4} Q_{a,b}^2 + \frac{N\beta^2}{2} q_N(\sigma^{(a)},\sigma^{(b)})\, Q_{a,b}} \tag{1.58}$$

and imposing the symmetry $Q_{a,b} = Q_{b,a}$ as a consequence of the fact that $q_N(\sigma^{(a)}, \sigma^{(b)}) = q_N(\sigma^{(b)}, \sigma^{(a)})$, we can write

$$\mathbb{E}([\mathcal{Z}_N(\beta, h, J)]^n) = \int \prod_{a\neq b}\left(dQ_{a,b}\sqrt{\frac{\beta^2 N}{4\pi}}\right) \sum_{\{\sigma^{(1)},\dots,\sigma^{(n)}\}\in\Sigma_N^n}$$
$$\times\, e^{\beta h \sum_{i=1}^N \sum_{a=1}^n \sigma_i^{(a)} - \frac{\beta^2 N}{4}(\sum_{a\neq b} Q_{a,b}^2 - 1) + \frac{N\beta^2}{2}\sum_{a\neq b} q_N(\sigma^{(a)},\sigma^{(b)})\, Q_{a,b}}.$$

Using (1.7), which gives the expression for the spin overlap in terms of spin configurations, we can further rewrite

$$\mathbb{E}([\mathcal{Z}_N(\beta,h,J)]^n) = \int \prod_{a\neq b}\left(dQ_{a,b}\sqrt{\frac{\beta^2 N}{4\pi}}\right) \sum_{\{\sigma^{(1)},\dots,\sigma^{(n)}\}\in\Sigma_N^n}$$

$$\times\, e^{\sum_{i=1}^N\left(\beta h\sum_{a=1}^n \sigma_i^{(a)} - \frac{\beta^2}{4}(\sum_{a\neq b}Q_{a,b}^2 - 1) + \frac{\beta^2}{2}\sum_{a\neq b}\sigma_i^{(a)}\sigma_i^{(b)}Q_{a,b}\right)}.$$

The expression above can be factorized over the sites and denoting by $S = (S_1,\dots,S_n)$ the generic n-fold replica of a spin at site $i\in\{1,\dots,N\}$ we obtain

$$\mathbb{E}([\mathcal{Z}_N(\beta,h,J)]^n) = \int \prod_{a\neq b}\left(dQ_{a,b}\sqrt{\frac{\beta^2 N}{4\pi}}\right) e^{-\frac{N\beta^2}{4}(\sum_{a\neq b}Q_{a,b}^2 - 1)}$$

$$\times\left(\sum_{S_1,\dots,S_n\in\Sigma_n} e^{\beta h\sum_{a=1}^n S_a + \frac{\beta^2}{2}\sum_{a\neq b}S_a S_b Q_{a,b}}\right)^N.$$

Finally, the use of the saddle-point method in the limit of large N and the imposition of the constraint $Q_{a,a}=1$ allow $p(\beta,h,n)$ to be re-written as an optimization problem of the function $A(Q,\beta,h,n)$ over the $n\times n$ matrix Q, as in Eq. (1.54). □

1.6 The replica symmetric solution

For integer $n\geqslant 1$ the stationary condition of the function $A(Q,\beta,h,n)$ in Eq. (1.55) corresponds to the $n(n-1)/2$ saddle-point equations

$$\frac{\partial A}{\partial Q_{a,b}} = 0 \qquad 1\leqslant a<b\leqslant n. \tag{1.59}$$

Note that the previous equations say that the $Q_{a,b}$ are the expectations of $S_a S_b$ with respect to the Boltzmann–Gibbs measure of Hamiltonian

$$H(S,Q,\beta,h) = \beta h\sum_{a=1}^n S_a + \frac{\beta^2}{2}\sum_{a,b=1}^n S_a S_b\, Q_{a,b}.$$

To finish the computation of the n-dependent pressure $p(\beta,h,n)$ and then take the limit $n\to 0$ we need to solve the saddle-point equations (1.59). The function $A(Q,\beta,h,n)$ is left invariant when we exchange the lines (or rows) of the matrix Q and therefore the group of permutations of n elements is a symmetry of the problem. The simplest solution of the saddle-point equation is of the form

$$Q_{a,b} = \begin{cases} 1 & \text{if } a = b \\ q & \text{if } a \neq b \end{cases} \tag{1.60}$$

for some $0 \leqslant q \leqslant 1$. This is called the *replica symmetric ansatz*. This form of the matrix Q is the only solution which is left invariant by the group of permutation of n elements, i.e. such that $Q_{a,b} = Q_{\pi(a),\pi(b)}$ for all permutations π of $\{1, 2, \ldots, n\}$. With this choice, from the replica method, one obtains the following expression for the pressure, known as the replica-symmetric solution, first found in Sherrington and Kirkpatrick (1975).

Replica computation 2 (Replica symmetric pressure) *Let us denote $d\mu(z) = \frac{1}{\sqrt{2\pi}} e^{-z^2/2} dz$. The replica symmetric expression for the pressure of the SK model is given by*

$$p^{(RS)}(\beta, h) = \frac{\beta^2}{4}(1-q)^2 + \ln 2 + \int d\mu(z) \ln\cosh(\beta[h + \sqrt{q}z]) \tag{1.61}$$

where $q = q(\beta, h)$ is the solution of the equation

$$q = \int d\mu(z) \tanh^2(\beta[h + \sqrt{q}z]). \tag{1.62}$$

Proof To reproduce the replica symmetric pressure (1.61) within the replica approach, let us compute the value of the n-dependent pressure of the SK model when the matrix Q is chosen as in Eq. (1.60). We then have

$$\sum_{a,b=1}^{n} Q_{a,b}^2 = n(n-1)q^2 + n\,. \tag{1.63}$$

Moreover,

$$\sum_{S_1,\ldots,S_n \in \Sigma_n} e^{\beta h \sum_{a=1}^n S_a + \frac{\beta^2}{2} \sum_{a,b=1}^n S_a S_b Q_{a,b}} = \sum_{S_1,\ldots,S_n \in \Sigma_n} e^{\beta h \sum_{a=1}^n S_a + \frac{\beta^2}{2} q (\sum_{a=1}^n S_a)^2 + \frac{\beta^2}{2} n(1-q)} \tag{1.64}$$

and applying the Hubbard–Stratonovich identity in the last formula, we decouple the spins $\{S_a\}$ and arrive at

$$\sum_{S_1,\ldots,S_n \in \Sigma_n} e^{\beta h \sum_{a=1}^n S_a + \frac{\beta^2}{2} \sum_{a,b=1}^n S_a S_b Q_{a,b}} = e^{\frac{\beta^2}{2} n(1-q)} \int d\mu(z) 2^n \cosh^n(\beta(h + z\sqrt{q})). \tag{1.65}$$

Therefore, using (1.63) and (1.65), the replica symmetric expression of (1.55) reads

$$A^{(RS)}(q,\beta,h,n) = \frac{\beta^2}{4}(1-2q-(n-1)q^2)+\ln 2+\frac{1}{n}\ln\int d\mu(z)\cosh^n(\beta[h+\sqrt{q}z]). \quad (1.66)$$

The stationary condition $\frac{\partial A^{(RS)}}{\partial q}=0$, followed by integration by parts, implies that $q=q(\beta,h,n)$ is the solution of the equation

$$q=\frac{\int d\mu(z)\cosh^n(\beta[h+\sqrt{q}z])\tanh^2(\beta[h+\sqrt{q}z])}{\int d\mu(z)\cosh^n(\beta[h+\sqrt{q}z])}. \quad (1.67)$$

Taking the limit $n\to 0$ in (1.66) and (1.67) we obtain the replica symmetric expression for the pressure. □

At zero magnetic field $h=0$, Eq. (1.62) only has the solution $q=0$ for $\beta<1$. For $\beta>1$ there are two other symmetric solutions where q is different from zero (they are supposed to be the physical ones with lower free energy). Therefore, there is a phase transition in the replica symmetric approach at the critical value $\beta_c=1$ for $h=0$, while there is no transition for $h\neq 0$.

We have the impression of being in a situation very similar to the one of the Curie–Weiss model, with the overlap q of Eq. (1.62) playing for the SK model the role of the magnetization m of Eq. (1.31) for the Curie–Weiss model. However, this belief is wrong. The phase transition of the SK model is much richer than the ferromagnetic transition. Moreover, in the plane (β,h) the replica method found that the replica symmetric solution is correct only in a region delimited by the so-called de Almeida–Thouless line (in particular, at $h=0$ the replica solution is correct only for temperature values above the critical temperature), as specified by the following result.

Replica computation 3 (de Almeida–Thouless line) *The thermodynamic limit of the pressure of SK mode coincides with the replica symmetric expressions (1.61) and (1.62) only in the region*

$$\beta^2\int d\mu(z)\frac{1}{\cosh^4(\beta[h+\sqrt{q}z])}\leqslant 1. \quad (1.68)$$

For rigorous analysis of the de Almeida–Thouless line, see Talagrand (2010b), Guerra (2006), Toninelli (2002). The fact that the replica symmetric solution is wrong at $h=0$ for temperature values below the critical temperature is reflected in the fact that at very low temperatures the thermodynamic quantities computed from expression (1.61) become unphysical. For example, the replica symmetric entropy at zero temperature turns out to be negative. This is not acceptable because the entropy is the logarithm of the number of configurations and in a discrete system there is always at least one state available and thus the entropy must be non-negative. The correct solution was found by Parisi with his ingenious ansatz of replica symmetry breaking.

1.7 The ultrametric replica symmetry breaking solution

Parisi made an hierarchical ansatz for the elements of the matrix Q. This is described by the following construction, which encodes ultrametricity.

One starts with the full matrix in which all elements of the symmetric $n \times n$ matrix Q have the same value $0 \leqslant q_1 \leqslant 1$ (except the diagonal ones which are taken to be one), as in the replica symmetric ansatz. Defining $m_1 = n$, one then breaks the $m_1 \times m_1$ matrix into $m_1/m_2 \times m_1/m_2$ blocks of size $m_2 \times m_2$. In the off-diagonal blocks one leaves q_1, but in the blocks on the diagonal one replaces q_1 by q_2 (with $1 \geqslant q_2 \geqslant q_1 \geqslant 0$). One then does the same thing within each of the blocks along the diagonal: they are broken into $m_2/m_3 \times m_2/m_3$ sub-blocks, each of size $m_3 \times m_3$, and in those along the diagonal one replaces q_2 by q_3 (with $1 \geqslant q_3 \geqslant q_2 \geqslant q_1 \geqslant 0$). This procedure is repeated in a nested way R times, with R an integer, and defines the R-steps replica symmetry breaking scheme.

To really be implemented, the R-steps replica symmetry breaking construction requires that each m_i has to divisible by m_{i+1}, which in particular requires

$$n = m_1 \geqslant m_2 \geqslant m_3 \geqslant \cdots \geqslant m_{R+1} \geqslant m_{R+2} = 1. \tag{1.69}$$

Under such assumptions of an ultrametric form of the matrix Q (specified by an admissible sequence of integers $\{m_i\}$ and a non-decreasing sequence $\{q_i\}$), Parisi suggested to work out the expression for the function $A(Q, \beta, h, n)$ in Eq. (1.55). This can be done by applying, for each nested block, the Hubbard–Stratonovich transformation to linearize the term in the exponential which is quadratic in the sum of the spin variables of the block. Parisi argued that the pressure $p(\beta, h)$ of the SK model can be calculated by considering a hierarchical scheme with an infinite number of steps and then taking the limit $n \to 0$. To this aim he made a number of assumptions. For instance, he claimed that in the limit $n \to 0$ one must turn the above sequence around and take

$$0 = m_1 \leqslant m_2 \leqslant m_3 \leqslant \cdots \leqslant m_{R+1} \leqslant m_{R+2} = 1. \tag{1.70}$$

Moreover, Parisi argued that in the region $0 < n < 1$ instead of maximizing the function $A(Q, \beta, h, n)$, as in the natural prescription of (1.54), one should minimize it.

As exotic as the ultrametric choice might be, and as strange as it might seem to deal with the elements of a $n \times n$ matrix in the limit $n \to 0$, nowadays we have a rigorous proof, due to Guerra and Talagrand, that the Parisi scheme of replica symmetry breaking gives the exact solution of the pressure of the SK model (we will present some of those rigorous results in Chapter 4).

To state the definite solution, with an infinite number of steps of replica symmetry breaking, we will first analyze the case of 1-step replica symmetry breaking (the simplest case of ultrametric ansatz) and then the R-step with integer $R > 1$.

Replica computation 4 (1-step replica symmetry breaking pressure) *The 1-step replica symmetry breaking (RSB) expression for the pressure of the SK model is given by*

$$p^{(1-\mathrm{RSB})}(\beta, h)$$
$$= \inf_{\substack{0\leqslant q_1\leqslant q_2\leqslant 1,\\ 0\leqslant m\leqslant 1}} \left[\frac{\beta^2}{4}(1+mq_1^2+(1-m)q_2^2-2q_2)+\ln 2 \right.$$
$$\left. + \int d\mu(z_1)\ln\left(\int d\mu(z_2)\cosh^m(\beta[h+\sqrt{q_2-q_1}z_2+\sqrt{q_1}z_1])\right)^{1/m}\right]. \tag{1.71}$$

Proof To obtain the 1-step replica symmetry breaking expression (1.71) within the replica method, one has to plug the Parisi ultrametric ansatz for $R = 1$ into (1.55). Namely, all the $n = m_1$ replicas are divided into n/m groups of size $m = m_2$. Therefore the matrix Q has the following elements

$$Q_{a,b} = \begin{cases} 1 & \text{if } a = b, \\ q_2 & \text{if } \{a \neq b\} \text{ belong to the same group of size } m, \\ q_1 & \text{if } \{a \neq b\} \text{ do not belong to the same group of size } m. \end{cases} \tag{1.72}$$

With this parametrization of the matrix Q, for the first term in Eq. (1.55) one has

$$-\frac{\beta^2}{4n}\sum_{a,b=1}^{n} Q_{a,b}^2 = -\frac{\beta^2}{4n}\left(n+q_2^2(m^2-m)\frac{n}{m}+q_1^2\left(n^2-m^2\frac{n}{m}\right)\right)$$
$$= -\frac{\beta^2}{4}\left((1-q_2^2)+m(q_2^2-q_1^2)+nq_1^2\right). \tag{1.73}$$

To evaluate the second term in Eq. (1.55)

$$\frac{1}{n}\ln\left(\sum_{\{S_1,\ldots,S_n\}\in\Sigma_n} e^{\beta h\sum_{a=1}^n S_a+\frac{\beta^2}{2}\sum_{a,b=1}^n S_aS_bQ_{a,b}}\right), \tag{1.74}$$

we write

$$\sum_{a,b=1}^{n} Q_{a,b}S_aS_b = q_1\left(\sum_{a=1}^{n} S_a\right)^2 + (q_2-q_1)\sum_{i=1}^{n/m}\left(\sum_{a=1}^{m} S_{i,a}\right)^2 + (1-q_2)n. \tag{1.75}$$

where $S_{i,a}$ denotes the ath spin (with $a = 1, \ldots, m$) in the ith block (with $i = 1, \ldots, n/m$). We also introduce independent standard Gaussian random variables $Z_1, Z_{2,1}, \ldots, Z_{2,n/m}$ to linearize the quadratic terms in the exponent using a series of Hubbard–Stratonovich transformations, so that (1.74) becomes

$$\frac{\beta^2}{2}(1-q_2) + \frac{1}{n}\ln\left(\sum_{\{S_1,\ldots,S_n\}\in\Sigma_n} e^{\beta h\sum_{a=1}^{n} S_a}\mathbb{E}_{Z_1}\left(e^{\beta\sqrt{q_1}\left(\sum_{a=1}^{m} S_a\right)Z_1}\right.\right.$$
$$\left.\left.\times\ \mathbb{E}_{Z_{2,1},\ldots,Z_{2,n/m}}\left(e^{\beta\sqrt{q_2-q_1}\sum_{i=1}^{n/m}\left(\sum_{a=1}^{n} S_{i,a}\right)Z_{2,a}}\right)\right)\right), \tag{1.76}$$

where $\mathbb{E}_Z$ denotes expectation with respect to the random variables Z. Performing the sum over the spin, one obtains

$$\frac{\beta^2}{2}(1-q_2) + \frac{1}{n}\log\left(\mathbb{E}_{Z_1}((\mathbb{E}_{Z_2}(\cosh^m(\beta[h+\sqrt{q_1}Z_1+\sqrt{q_2-q_1}Z_2])))^{n/m})\right). \tag{1.77}$$

Inserting (1.73) and (1.77) into (1.55), and taking the limit $n \to 0$, the 1-step replica symmetric expression (1.71) is recovered. This also requires the unjustified assumptions of changing the sup to an inf in (1.54) and taking $0 \leqslant m \leqslant 1$. □

Remark 1.15 Defining $m_1 = 0, m_2 = m, m_3 = 1$, and $q_0 = 0, q_3 = 1$, the 1-step replica symmetry breaking expression of the pressure (1.71) can be rewritten as

$$p^{(1-\mathrm{RSB})}(\beta,h)$$
$$= \inf_{\substack{0=q_0\leqslant q_1\leqslant q_2\leqslant q_3=1\\ 0=m_1\leqslant m_2\leqslant m_3=1}}\left[-\frac{\beta^2}{4}\left(m_1(q_1^2-q_0^2)+m_2(q_2^2-q_1^2)+m_3(q_3^2-q_2^2)\right)+\ln 2\right.$$
$$+\ln\left(d\mu(z_1)\left(\int d\mu(z_2)\left(\int d\mu(z_3)\cosh^{m_3}(\beta[h+\sqrt{q_3-q_2}z_3\right.\right.\right.$$
$$\left.\left.\left.\left.+\sqrt{q_2-q_1}z_2+\sqrt{q_1-q_0}z_1])\right)^{\frac{m_2}{m_3}}\right)^{\frac{m_1}{m_2}}\right)^{\frac{1}{m_1}}\right]. \tag{1.78}$$

with the convention that setting the value $m_1 = 0$ in the expression above between square brackets corresponds to taking the limit $m_1 \to 0$ of the same expression.

The previous remark shows us the general form of the replica symmetry breaking solution for an arbitrary number of steps $R \in \{0, 1, 2, \ldots\}$. Defining $k = R + 2$, the pressure will be expressed as an optimization problem of a function similar to (1.78) but now involving two sequences

$$0 = q_0 \leqslant q_1 \leqslant \cdots \leqslant q_{k-1} \leqslant q_k = 1$$

and

$$0 = m_1 \leqslant m_2 \leqslant \cdots \leqslant m_{k-1} \leqslant m_k = 1. \tag{1.79}$$

Replica computation 5 (*R*-step replica symmetry breaking pressure) *The R-step replica symmetry breaking pressure of the SK model is given by*

$$p^{(R-\mathrm{RSB})}(\beta, h)$$
$$= \inf_{\substack{0=q_0\leqslant q_1\leqslant\ldots\leqslant q_{k-1}\leqslant q_k=1 \\ 0=m_1\leqslant m_2\leqslant\ldots\leqslant m_{k-1}\leqslant m_k=1}} \left[-\frac{\beta^2}{4}\sum_{l=1}^{k} m_l\left(q_l^2 - q_{l-1}^2\right) + \ln 2 + \log(Z_0) \right] \tag{1.80}$$

where $(Z_l)_{0\leqslant l\leqslant k}$ *is the succession which is defined by the final condition*

$$Z_k = \cosh\left[\beta\left(h + \sum_{l=1}^{k} X_l\sqrt{q_l - q_{l-1}}\right)\right] \tag{1.81}$$

and iteratively, for $1 \leqslant l \leqslant k$,

$$Z_{l-1} = \left(\mathbb{E}_l\left(Z_l^{m_l}\right)\right)^{\frac{1}{m_l}} \tag{1.82}$$

where $(X_l)_{1\leqslant l\leqslant k}$ *are i.i.d. standard Gaussian random variables and* $\mathbb{E}_l$ *denotes expectation with respect to the l-th of them. In (1.80) the convention is that setting the value* $m_1 = 0$ *in the expression between square brackets corresponds to taking the limit* $m_1 \to 0$ *in the same expression.*

Proof Expression (1.80) can be obtained by a computation analogous to one of Replica computation 4 (p. 20), starting now from an admissible sequence

$$n = m_1 \geqslant m_2 \geqslant m_3 \geqslant \cdots \geqslant m_{k-1} \geqslant m_k = 1 \tag{1.83}$$

and a parametrization of the matrix Q given by

$$Q_{a,b} = \begin{cases} q_k = 1 & \text{if } a = b, \\ q_{k-1} & \text{if } \{a \neq b\} \text{belong to the same group of size} m_{k-1}, \\ q_{k-2} & \text{if } \{a \neq b\} \text{ do not belong to the same group of size } m_{k-1} \\ & \quad \text{and they do belong to the same group of size } m_{k-2}, \\ \vdots & \\ \vdots & \\ q_1 & \text{if } \{a \neq b\} \text{ do not belong to the same group of size } m_{k-1} \\ & \quad \text{and they do not belong to the same group of size } m_{k-2}, \\ & \quad \ldots, \\ & \quad \text{and they do not belong to the same group of size } m_2, \\ & \quad \text{and they do belong to the same group of size } m_1. \end{cases} \tag{1.84}$$

Note once more that in the limit $n \to 0$, the inequalities in the sequence (1.83) must be reversed, giving rise to the sequence in (1.79).

We are now in the position to state the final expression of the Parisi solution for the pressure of the SK model in the limit of an infinite number of steps of replica symmetry breaking. Consider the set $\mathcal{M}$ of non-decreasing functions on the unit interval:

$$\mathcal{M} = \{x(q) : [0, 1] \to [0, 1], \quad x(q) \text{ non-decreasing}\}. \qquad \square$$

Definition 1.16 (Parisi pressure) For a given inverse temperature β and external field h, the Parisi pressure is given by the minimization of the Parisi functional, i.e.

$$p^{(Parisi)}(\beta, h) = \inf_{x(q) \in \mathcal{M}} \mathcal{P}(x(q), \beta, h) \tag{1.85}$$

where, for a given $x(q) \in \mathcal{M}$,

$$\mathcal{P}(x(q), \beta, h) = \log 2 + f(0, h) - \frac{\beta^2}{2} \int_0^1 q\, x(q) dq \tag{1.86}$$

and where $f(q, y) : [0, 1] \times \mathbb{R} \to \mathbb{R}$ is the function which satisfies the Parisi partial differential equation

$$\frac{\partial f}{\partial q} + \frac{1}{2}\frac{\partial^2 f}{\partial y^2} + \frac{1}{2} x(q) \left(\frac{\partial f}{\partial y}\right)^2 = 0, \tag{1.87}$$

with initial data

$$f(1, y) = \log \cosh(\beta y). \tag{1.88}$$

We shall show that in the case of a piecewise constant $x(q)$, the Parisi expression (1.85) yields the R-step replica symmetry breaking pressure (1.80). Since any non-decreasing function of the unit interval can be approximated by a piecewise constant order parameter, this shows how the Parisi expression (1.85) was obtained in the context of the replica method.

To analyze the case of a piecewise constant $x(q)$ the following result will be useful.

Lemma 1.17 (Solution Parisi PDE, constant $x(q)$) *For* $0 \leqslant m \leqslant 1$ *and* $f(q, y) : [q_a, q_b] \times \mathbb{R} \to \mathbb{R}$, *consider the partial differential equation*

$$\frac{\partial f}{\partial q} + \frac{1}{2}\frac{\partial^2 f}{\partial y^2} + \frac{1}{2} m \left(\frac{\partial f}{\partial y}\right)^2 = 0 \tag{1.89}$$

with final condition

$$f(q_b, y) = g(y). \tag{1.90}$$

Then the solution is given by:

- *if* $m = 0$ *then*

$$f(q, y) = \mathbb{E}\left(g(y + X\sqrt{q_b - q}\,)\right) \tag{1.91}$$

- *if* $m \neq 0$ *then*

$$f(q, y) = \log\left(\mathbb{E}\left(e^{mg(y+X\sqrt{q_b-q})}\right)\right)^{\frac{1}{m}} \tag{1.92}$$

where $\mathbb{E}$ *denotes expectation with respect to a standard Gaussian random variable* X.

Proof of Lemma 1.17 The case $m = 0$ is immediately solved because the equation loses its nonlinear part and it becomes the backward heat equation

$$\frac{\partial f}{\partial q} + \frac{1}{2}\frac{\partial^2 f}{\partial y^2} = 0 \tag{1.93}$$

which has the solution, for q in the interval $[q_a, q_b)$,

$$f(q, y) = \int dx f(q_b, x) \frac{e^{-\frac{1}{2}\frac{(x-y)^2}{(q_b-q)}}}{\sqrt{2\pi(q_b - q)}}.$$

The final condition (1.90) and the substitution $z = (x - y)/\sqrt{q_b - q}$ yields (1.91).

The case $m \neq 0$ is linearized through the substitution

$$\tilde{f}(q, y) = e^{mf(q,y)}.$$

It turns out that since f satisfies equation (1.89), then $\tilde{f}$ satisfies the backward heat equation (1.93) and therefore

$$\tilde{f}(q, y) = \int dx\, \tilde{f}(q_b, x) \frac{e^{-\frac{1}{2}\frac{(x-y)^2}{(q_b-q)}}}{\sqrt{2\pi(q_b - q)}},$$

which is equivalent to

$$f(q, y) = \frac{1}{m} \log \left(\int dx e^{m f(q_b, x)} \frac{e^{-\frac{1}{2}\frac{(x-y)^2}{(q_b-q)}}}{\sqrt{2\pi(q_b - q)}} \right).$$

As before, the final condition (1.90) and the substitution $z = (x - y)/\sqrt{q_b - q}$ yields (1.92). □

Armed with the previous lemma, it is now easy to obtain an explicit expression of the Parisi functional for piecewise constant order parameters $x(q)$. For a given $k = R + 2$ one considers the two sequences

$$0 = q_0 \leqslant q_1 \leqslant \cdots \leqslant q_k = 1$$

and

$$0 = m_1 \leqslant m_2 \leqslant \cdots \leqslant m_{k-1} \leqslant m_k = 1, \tag{1.94}$$

and the piecewise constant order parameter given by

$$x(q) = \begin{cases} m_1 = 0 & \text{if } 0 = q_0 \leqslant q < q_1 \\ m_2 & \text{if } q_1 \leqslant q < q_2 \\ \vdots & \\ m_{k-1} & \text{if } q_{k-2} \leqslant q < q_{k-1} \\ m_k = 1 & \text{if } q_{k-1} \leqslant q \leqslant q_k = 1. \end{cases} \tag{1.95}$$

Inside each interval the order parameter is constant and consequently the two terms in the Parisi functional (1.86) take the following expression:

$$\frac{\beta^2}{2} \int_0^1 q\, x(q) dq = \frac{\beta^2}{4} \sum_{l=1}^{k} m_l \left(q_l^2 - q_{l-1}^2\right) \tag{1.96}$$

and

$$f(0, h) = \log(Z_0) \tag{1.97}$$

where Z_0 is defined in (1.81) and (1.82).

Indeed, the first equation (1.96) follows immediately from the insertion of the piecewise constant order parameter (1.95), followed by elementary integration.

The second equation (1.97) is obtained by using the fact that $f(0, h)$ is the solution $f(q, y)$ of the Parisi partial differential equation (1.87) evaluated at $q = 0$ and $y = h$. In the case of a piecewise constant order parameter, this solution can be obtained iteratively by using the result of Lemma 1.17. One starts from the final condition in the Parisi PDE

$$f(1, y) = \log\cosh(\beta y).$$

In the interval $q_{k-1} \leqslant q < q_k = 1$ the order parameter equals m_k. Therefore, in this interval, Lemma 1.17 gives

$$\begin{aligned} f(q, y) &= \log\left(\mathbb{E}_k\left(e^{m_k \log\cosh[\beta(y + X_k\sqrt{q_k - q})]}\right)\right)^{\frac{1}{m_k}} \\ &= \log\left(\mathbb{E}_k\left(\left\{\cosh\left[\beta\left(y + X_k\sqrt{q_k - q}\right)\right]\right\}^{m_k}\right)\right)^{\frac{1}{m_k}}, \end{aligned}$$

and at the left extreme of the interval one has

$$f(q_{k-1}, y) = \log\left(\mathbb{E}_k\left(\left\{\cosh\left[\beta\left(y + X_k\sqrt{q_k - q_{k-1}}\right)\right]\right\}^{m_k}\right)\right)^{\frac{1}{m_k}}.$$

Then one proceeds to the the interval $q_{k-2} \leqslant q < q_{k-1}$ where the order parameter equals m_{k-1}. Applying Lemma 1.17 again, with the final condition above for $f(q_{k-1}, y)$, gives

$$\begin{aligned} f(q, y) = \log\Bigg(\mathbb{E}_{k-1}\Big(\mathbb{E}_k\Big(\Big\{\cosh\Big[\beta\Big(y + X_k\sqrt{q_k - q_{k-1}} \\ + X_{k-1}\sqrt{q_{k-1} - q}\,\Big)\Big]\Big\}^{m_k}\Big)\Big)^{\frac{m_{k-1}}{m_k}}\Bigg)^{\frac{1}{m_{k-1}}}. \end{aligned}$$

Continuing iteratively in this way for all intervals $q_{l-1} \leqslant q < q_l$ until $l = 1$ and substituting $y = h$ one finally obtains (1.97). □

The Parisi solution of the SK model is based on an ultrametric ansatz for the matrix Q and the analytical continuation $n \to 0$. Even though it was immediately accepted by the theoretical physics community as the correct solution of the SK model (based on the evidence that it was in very good agreement with the results of numerical simulations), for 30 years it remained a mystery not accessible to rigorous analysis. It was only in recent years that the expression (1.85) was finally shown to be the true pressure of the SK model in a rigorous self-contained mathematical approach which does not involve any replica computation. We will return to those results in Chapter 4.

2
Correlation inequalities

Abstract

In this chapter we present a number of correlation inequalities for spin systems. We start by reproducing the classical Griffiths–Kelly–Sherman inequalities for ferromagnetic spin models by means of the Ginibre method. We then show how to prove a type I correlation inequality for spin glasses both in the Gaussian case and for general centered coupling distributions. The extension to the quantum case is also treated. Type II inequalities are discussed and proved for one-dimensional systems and counterexamples are shown above one dimension. A complete inequalities theory is presented for Gaussian spin glass models on the Nishimori line.

2.1 Spin functions

In a set Λ of cardinality $|\Lambda| = N$ the subsets $X \subset \Lambda$ belong to the set of parts Γ_Λ (which includes the empty set $\emptyset$ and the set Λ itself). The set of spin configurations in the volume Λ called Σ_Λ has the same cardinality as Γ_Λ:

$$|\Gamma_\Lambda| = |\Sigma_\Lambda| = 2^N. \tag{2.1}$$

The two sets are moreover isomorphic in the sense that there exists a one-to-one map which preserves the group structure that both Σ_Λ and Γ_Λ are endowed with. The one-to-one correspondence associates the subset $X \in \Gamma_\Lambda$ with the spin configuration $\sigma \in \Sigma_\Lambda$ whose minus signs are localized in X:

$$\sigma_i = \begin{cases} -1 & \text{if } i \in X, \\ +1 & \text{if } i \notin X. \end{cases} \tag{2.2}$$

The group structures of Σ_Λ and Γ_Λ are obtained as follows. The first is endowed with the pointwise multiplication: given $\sigma = \{\sigma_i\}_{i\in\Lambda}$ and $\tau = \{\tau_i\}_{i\in\Lambda}$ the product is defined by $\sigma\tau = \{\sigma_i\tau_i\}_{i\in\Lambda}$. The second is endowed with the symmetric difference product: given X and Y the product is $XY = (X \cup Y) \setminus (X \cap Y)$. The identity in Σ_Λ is the configuration with all spins equal to 1, while the identity in Γ_Λ is the empty set. We notice the two groups have squares equal to the identities $\sigma_i^2 = 1$ and $X^2 = \emptyset$. Thus in the sequel functions of spin configurations, σ will be equivalently thought of as functions of subsets X.

Important examples of functions from Γ_Λ to $\mathbb{Z}_2$ are the 2^N functions σ_X defined by

$$\sigma_X(Y) = (-1)^{|X\cap Y|}. \tag{2.3}$$

One can see from inspection that

$$\sigma_X(Y) = \sigma_Y(X). \tag{2.4}$$

The definition of the same function on the isomorphic set Σ_Λ is given by

$$\sigma_X = \prod_{i\in X} \sigma_i, \tag{2.5}$$

where

$$\sigma_i(\tau) = \begin{cases} +1 & \text{if } \tau_i = +1, \\ -1 & \text{if } \tau_i = -1. \end{cases} \tag{2.6}$$

Clearly

$$\sigma_X\sigma_Y = \sigma_{XY}. \tag{2.7}$$

Applying (2.3), one can immediately verify that they have the remarkable *character* property

$$\sigma_X(A)\sigma_X(B) = \sigma_X(AB). \tag{2.8}$$

The general functions from Γ_Λ to $\mathbb{R}$ form a vector space endowed with the scalar product

$$\langle f, g\rangle = 2^{-N} \sum_{X\subset\Lambda} f(X)g(X); \tag{2.9}$$

the scalar product corresponds to the expectation, with respect to the uniform probability measure $\omega^{(0)}$ on spin configurations, of the product of the two functions

f and g:

$$\langle f, g \rangle = \omega^{(0)}(fg) = 2^{-N} \sum_{\sigma \in \Sigma_\Lambda} f(\sigma) g(\sigma). \tag{2.10}$$

With respect to the previous scalar product, the characters form an orthonormal system of vectors. The orthonormality can be proved by observing that the local spins σ_i are independent random variables w.r.t. $\omega^{(0)}$, i.e.

$$\omega^{(0)}(\sigma_X) = \prod_{i \in X} \omega^{(0)}(\sigma_i) = \delta_{X,\emptyset}. \tag{2.11}$$

By consequence

$$\langle \sigma_X, \sigma_Y \rangle = \omega^{(0)}(\sigma_X \sigma_Y) = \omega^{(0)}(\sigma_{XY}) = \delta_{XY,\emptyset} = \delta_{X,Y}. \tag{2.12}$$

This immediately leads to the following.

Theorem 2.1 (Spin functions) *For each function $f : \Gamma_\Lambda \to \mathbb{R}$ there exists a set of coefficients labeled by X which can be computed as $\langle f, \sigma_X \rangle$ s.t.*

$$f = \sum_{X \subset \Lambda} \langle f, \sigma_X \rangle \sigma_X. \tag{2.13}$$

Proof This follows from the fact that the characters of (2.5) form an orthonormal base. □

Remark 2.2 From the algebraic point of view, the previous theorem corresponds to the computation of the Fourier transform of f:

$$\widehat{f} = \sum_{X \subset \Lambda} f(X) \sigma_X, \tag{2.14}$$

and its inverse

$$f = 2^{-N} \sum_{X \subset \Lambda} \widehat{f}(X) \sigma_X. \tag{2.15}$$

Based on the previous theorem we will describe physical spin systems by the general Hamiltonian

$$H = - \sum_{X \subset \Lambda} J_X \sigma_X, \tag{2.16}$$

where the coefficients J_X play the role of multi-interaction, namely magnetic fields when $|X| = 1$; two-body interactions when $|X| = 2$; k-body interactions when

$|X| = k$. The empty set coefficient $J_\emptyset$ does not play any role since it cancels in the Boltzmann–Gibbs measure.

We also observe that from (2.14) and (2.16), it follows that

$$\widehat{J} = -H. \tag{2.17}$$

The previous representation can be extended to random spin functions. One then obtains the following representation theorem for Gaussian processes labeled by spin configurations.

Theorem 2.3 (Gaussian processes on spin configurations) *Let H be a Gaussian family of centered random variables defined by the covariance $\mathcal{C}$: $\Gamma_\Lambda \times \Gamma_\Lambda \to \mathbb{R}$. There exists a family of independent centered Gaussians J_X of variance Δ_X^2 s.t.*

$$H = -\sum_{X \subset \Lambda} J_X \sigma_X, \tag{2.18}$$

where

$$\Delta^2 = \frac{1}{2^{2N}} \sum_{X,Y \subset \Lambda \times \Lambda} \mathcal{C}(X, Y) \sigma_X \tau_Y. \tag{2.19}$$

Proof In order to derive the quantities Δ_X^2, starting from the function $\mathcal{C}$ we observe that each function f can be written, in analogy with (2.13), as

$$f = \sum_{X,Y \subset \Lambda \times \Lambda} \langle f, \sigma_X \tau_Y \rangle \sigma_X \tau_Y \tag{2.20}$$

since $\sigma_X \tau_Y$ is an orthonormal system with respect to the scalar product

$$\langle f, g \rangle = 2^{-2N} \sum_{X,Y \subset \Lambda \times \Lambda} f(X, Y) g(X, Y). \tag{2.21}$$

Since from (2.18) we have

$$\mathcal{C} = \sum_{X \subset \Lambda} \Delta_X^2 \sigma_X \tau_X = \sum_{X,Y \subset \Lambda \times \Lambda} \Delta_X^2 \delta_{X,Y} \sigma_X \tau_Y \tag{2.22}$$

the use of the Fourier inverse transform (extended to functions on $\Gamma_\Lambda \times \Gamma_\Lambda$) provides the proof. □

Remark 2.4 From a geometrical point of view, the representation theorem that we just proved for a Gaussian family in terms of independent Gaussians corresponds to the invariance of a multivariate Gaussian measure with respect to rotations in R^{2^N}.

2.2 Ferromagnetism and the Griffiths–Kelly–Sherman inequalities

In a typical ferromagnetic system, two spins σ_1 and σ_2 interact with each other with a two body potential

$$-J_{1,2}\sigma_1\sigma_2 \tag{2.23}$$

where $J_{1,2} \geqslant 0$. The positivity of the parameter $J_{1,2}$ favors the alignment of the spins in the equilibrium Boltzmann–Gibbs state. Early studies of the spin models showed that ferromagnetic spin models should verify the property

$$\omega(\sigma_1\sigma_2) \geqslant 0, \tag{2.24}$$

where $\omega(\cdot)$ denotes expectation with respect to the Boltzmann–Gibbs measure. The previous inequality, called type I, was proved in the two-body case by Griffiths (1967a) together with another inequality called type II in Griffiths (1967b), and later generalized by Kelly and Sherman (1968). The simple proof that we propose here is due to Ginibre (1970).

We recall the standard definition for a general spin system with deterministic couplings.

Definition 2.5 (Boltzmann–Gibbs measure, deterministic case) For a given Hamiltonian of the form

$$H_\Lambda(\sigma) = -\sum_{X\subset\Lambda} J_X\sigma_X \tag{2.25}$$

with deterministic couplings $J = \{J_X\}$ and a spin function $f\colon \Sigma_\Lambda \to \mathbb{R}$, the Boltzmann–Gibbs expectation of f is given by

$$\omega(f) = \frac{\sum_{\sigma\in\Sigma_\Lambda} f(\sigma)\exp\left[-H_\Lambda(\sigma)\right]}{\sum_{\sigma\in\Sigma_\Lambda}\exp\left[-H_\Lambda(\sigma)\right]}. \tag{2.26}$$

The quantity

$$Z_\Lambda = \sum_{\sigma\in\Sigma_\Lambda}\exp\left[-H_\Lambda(\sigma)\right] \tag{2.27}$$

is called *partition function*, and

$$P_\Lambda = \log Z_\Lambda \tag{2.28}$$

is the *pressure*.

Remark 2.6 The usual dependence on the inverse temperature β is absorbed here in the parameters J. We will make it explicit when needed.

Remark 2.7 We refer here to *free boundary conditions* on general topologies (cubes, parallelepipeds, tori, etc.), i.e. with a partition function summed over all spin configurations, with no restrictions.

Introducing the convolution product:

$$f * g(X) = \sum_{Y \subset X} f(Y) g(XY) \tag{2.29}$$

one can prove, just by using the definitions from the previous section, that

$$\widehat{f}\,\widehat{g} = \widehat{f * g} \tag{2.30}$$

i.e. the Fourier transform maps the convolution product into the ordinary product. In particular, indicating with f^{n*} the nth convolution power

$$(\hat{f})^n = \widehat{f^{n*}}, \tag{2.31}$$

which implies

$$e^{\widehat{f}} = \widehat{\exp^{*f}}, \tag{2.32}$$

where $\exp^{*f} = \sum_{n=0}^{\infty} \frac{1}{n!} f^{n*}$. With this setup, one can prove the following result.

Theorem 2.8 (Griffiths–Kelly–Sherman inequalities) *For the Hamiltonian (2.25) with $J_X \geqslant 0 \,\forall\, X \subset \Lambda$ the following two inequalities hold*

$$\frac{\partial P_\Lambda}{\partial J_X} = \omega(\sigma_X) \geqslant 0, \tag{2.33}$$

$$\frac{\partial^2 P_\Lambda}{\partial J_X \partial J_Y} = \omega(\sigma_{XY}) - \omega(\sigma_X)\omega(\sigma_Y) \geqslant 0. \tag{2.34}$$

Remark 2.9 The hypothesis of the Griffiths–Kelly–Sherman inequalities excludes those boundary conditions which are equivalent to negative values of some of the J's – for instance, the "minus" boundary conditions, where all the spins on the boundary of the volume Λ are fixed at -1, are equivalent to having a $-\infty$ field on these sites.

Proof Using (2.17) one has

$$2^{-N} Z \omega(\sigma_X) = \langle \sigma_X, \exp^{\widehat{J}} \rangle; \tag{2.35}$$

on the other hand

$$\langle \sigma_X, \widehat{\exp^{*J}} \rangle = \exp^{*J}(X) \tag{2.36}$$

where we have used the properties (2.4), (2.15). By (2.32) one has

$$2^{-N} Z\omega(\sigma_X) = \exp^{*J}(X). \tag{2.37}$$

Since

$$\exp^{*J}(X) = \sum_{n=0}^{\infty} \frac{1}{n!} \sum_{Y_1 Y_2 \cdots Y_n = X} J_{Y_1} J_{Y_2} \cdots J_{Y_n} \tag{2.38}$$

when the J_X are positive, the previous quantity is positive too, proving the general inequality of type I.

One may notice that the inequality just proved could have been directly deduced by a Taylor expansion of $Z \cdot \omega(\sigma_X)$ (see Glimm and Jaffe (1981)). Besides being useful to understand spin functions, the formalism introduced by Ginibre makes the proof of the second inequality much more transparent.

To prove the inequality of type II (2.34), one proceeds as follows:

$$\omega(\sigma_{XY}) - \omega(\sigma_X)\omega(\sigma_Y) = Z^{-2} \left(\sum_{A,B} [\sigma_{XY}(A) - \sigma_X(A)\sigma_Y(B)] e^{-H(A)-H(B)} \right). \tag{2.39}$$

Defining $AB = C$, $(B = CA)$ we get

$$\begin{aligned} \sigma_{XY}(A) - \sigma_X(A)\sigma_Y(B) &= \sigma_{XY}(A)[1 - \sigma_{XY}(A)\sigma_X(A)\sigma_Y(B)] \\ &= \sigma_{XY}(A)[1 - \sigma_Y(C)]. \end{aligned} \tag{2.40}$$

Observing that

$$\begin{aligned} -H(A) - H(B) &= \sum_Z J_Z(\sigma_Z(A) + \sigma_Z(B)) \\ &= \sum_Z J_Z \sigma_Z(A)[1 + \sigma_Z(C)] = \sum_Z \tilde{J}_Z \sigma_Z(A), \end{aligned} \tag{2.41}$$

where we have defined $\tilde{J}_Z(C) = J_Z[1 + \sigma_Z(C)] \geqslant 0$. Hence

$$\omega(\sigma_{XY}) - \omega(\sigma_X)\omega(\sigma_Y) = Z^{-2} \sum_C [1 - \sigma_Y(C)] \sum_A \sigma_{XY}(A) e^{-\sum_Z \tilde{J}_Z(C)\sigma_Z(A)} \tag{2.42}$$

which is positive because $[1 - \sigma_Y(C)] \geqslant 0$ and the sum in A is positive for (2.33). □

Types I and II inequalities have been extensively used to obtain many important results in statistical mechanics. They allow for instance the proof of the existence by monotonicity of the thermodynamic limit for pressure and correlation functions, using the first and second inequalities respectively. Moreover they have been used to prove bounds on thermodynamic quantities among different lattices, and bounds on critical temperatures and exponents (for an extended account, see Simon (1993)).

2.3 Spin glass correlation inequalities of type I

The previous section on ferromagnets shows that the pressure is a monotonic function with respect to the strength of the interaction. It is known that such a property is violated when the interactions are allowed to take negative signs. In this section we prove a correlation inequality for suitable quenched correlations which has a clear meaning: in a spin glass with centered interactions the pressure is monotonic with respect to the variance of the interaction. For the Gaussian interaction case we follow the proof by Contucci and Graffi (2004a) which uses integration by parts. The generalization to any centered random variable interaction is done following Contucci and Lebowitz (2007), where a simple convexity method is used to prove the result.

Theorem 2.10 (Type I inequality, centered Gaussian case) *For a spin glass defined by the Hamiltonian (1.8) with Gaussian-centered independent couplings $\{J_X\}$ with:*

$$\mathbb{E}\,[J_X] = 0, \tag{2.43}$$

and variance

$$\mathbb{E}\left[J_X^2\right] = \Delta_X^2, \tag{2.44}$$

the following inequality holds $\forall\ X \subset \Lambda$:

$$\langle J_X \sigma_X \rangle \ \geqslant 0. \tag{2.45}$$

Proof We remind the reader of the integration by parts for independent centered Gaussian variables J_X

$$\mathbb{E}\,[J_X f(J)] = \Delta_X^2 \mathbb{E}\left[\frac{df(J)}{dJ_X}\right], \tag{2.46}$$

and the correlation derivative formula

$$\frac{d\omega(\sigma_X)}{dJ_X} = 1 - \omega(\sigma_X)^2 \ \geqslant 0. \tag{2.47}$$

By successively applying (2.46) and (2.47) we obtain

$$\langle J_X \sigma_X \rangle = \mathbb{E}\,[J_X \omega(\sigma_X)] = \Delta_X^2 [1 - \mathbb{E}\left[\omega(\sigma_X)^2\right]] \geqslant 0. \tag{2.48}$$

□

Let us consider the Hamiltonian of type (1.8) expressed as

$$H_\Lambda(\sigma) = -\sum_{X\subset\Lambda} \lambda_X J_X \sigma_X, \tag{2.49}$$

where the $\lambda = \{\lambda_X\}_{X\subset\Lambda}$ are auxiliary parameters used to tune the strength of the interaction and the $J = \{J_X\}_{X\subset\Lambda}$ are independent centered random variables

$$\mathbb{E}[J_X] = 0. \tag{2.50}$$

Theorem 2.11 (Type I inequality, centered general case) *Let $\lambda_X \geq 0$. For the quenched measure induced by the spin glass Hamiltonian (2.49) with general centered independent couplings $\{J_X\}$ with $\mathbb{E}[J_X] = 0$, one has the inequality $\forall\, X \subset \Lambda$*

$$\frac{\partial P_\Lambda}{\partial \lambda_X} = \langle J_X \sigma_X \rangle \geqslant 0. \tag{2.51}$$

Proof The function P_Λ is convex with respect to λ_X, i.e.

$$\frac{\partial^2 P_\Lambda}{\partial \lambda_X^2} = \mathbb{E}\left[J_X^2[1-\omega^2(\sigma_X)]\right] \geqslant 0. \tag{2.52}$$

By consequence, the first derivative of the pressure is monotonically non-decreasing, i.e.

$$\frac{\partial P_\Lambda}{\partial \lambda_X}(\lambda_X^{(1)}) \geqslant \frac{\partial P_\Lambda}{\partial \lambda_X}(\lambda_X^{(2)}), \qquad \forall \lambda_X^{(1)} \geq \lambda_X^{(2)} \geq 0. \tag{2.53}$$

We now observe that when computed in $\lambda_X = 0$, the random variable $\omega(\sigma_X)$ is independent from the random variable J_X, thus implying

$$\frac{\partial P_\Lambda}{\partial \lambda_X}(\lambda_X = 0) = \mathbb{E}[J_X \omega(\sigma_X)]\,|_{\lambda_X=0} = \mathbb{E}[J_X]\,\mathbb{E}[\omega(\sigma_X)]\,|_{\lambda_X=0} = 0. \tag{2.54}$$

The theorem follows from (2.53) and (2.54). □

2.4 Extension to quantum case

Quantum spin systems with quenched randomness are both important and theoretically challenging. They are widely used as models for metallic alloys in condensed matter physics (see Sachdev (2001)) but also in combinatorial optimization problems – especially in relation to quantum annealing procedures (Morita and Nishimori (2008)) and quantum error correcting codes (Das and Chakrabarti (2005)). For the extension of type I inequalities to quantum systems we will be following Contucci and Lebowitz (2010).

For each finite set of points Λ, consider the quantum spin system with Hamiltonian

$$H_\Lambda = -\sum_{X\subset\Lambda} \lambda_X J_X \Phi_X. \tag{2.55}$$

The operators Φ_X are self-adjoint elements of the real algebra generated by the set of *spin operators*, the Pauli matrices, $\sigma_i^{(x)}, \sigma_i^{(y)}, \sigma_i^{(z)}$, $i \in \Lambda$, on the Hilbert space $\mathcal{H}_X := \otimes_{i\in X}\mathcal{H}_i$. The random interactions $\{J_X\}$ are centered, i.e. $\mathbb{E}[J_X] = 0$ for all X, and mutually independent. The $\lambda = \{\lambda_X\}_{X\subset\Lambda}$ are numbers which tune the magnitude of the random interactions.

Example 2.12 An example of a quantum spin glass model is the anisotropic quantum version of the nearest-neighbor Edwards–Anderson model with a transverse field. This is defined in terms of the Pauli matrices as

$$\begin{aligned}
\Phi_i &= \sigma_i^z, \\
\Phi_{i,j} &= \lambda_x \sigma_i^x \sigma_j^x + \lambda_y \sigma_i^y \sigma_j^y + \lambda_z \sigma_i^z \sigma_j^z, \qquad \text{for} \quad ||i-j|| = 1, \\
\Phi_X &= 0 \qquad\qquad\qquad\qquad\qquad\qquad\quad\;\; \text{for} \quad |X| > 2.
\end{aligned} \tag{2.56}$$

The main content of the previous section on classical spin glasses also holds true in the quantum setting. Namely, the pressure

$$P_\Lambda = \mathbb{E}\left[\log \operatorname{Tr} \exp(-H_\Lambda)\right], \tag{2.57}$$

is convex with respect to each λ_X. The proof of convexity is straightforward. The first derivative gives

$$\frac{\partial P_\Lambda}{\partial \lambda_X} = \mathbb{E}\left[J_X \omega(\Phi_X)\right] := \langle J_X \Phi_X \rangle, \tag{2.58}$$

where

$$\omega(f) = \frac{\operatorname{Tr} f e^{-H}}{\operatorname{Tr} e^{-H}}, \tag{2.59}$$

while, for the second derivative, one has (see Simon (1993), Chapter IV, page 357)

$$\frac{\partial^2 P_\Lambda}{\partial \lambda_X^2} = \mathbb{E}\left[J_X^2[(\Phi_X, \Phi_X) - \omega(\Phi_X)^2]\right] \tag{2.60}$$

where $(\cdot\,,\cdot)$ denotes the Duhamel inner product (see Dyson *et al.* (1978)):

$$(f, g) = \frac{\operatorname{Tr} \int_0^1 ds\, e^{-sH_\Lambda} f^* e^{-(1-s)H_\Lambda} g}{\operatorname{Tr} e^{-H_\Lambda}}. \tag{2.61}$$

By using the fact that $(1, g) = \omega(g)$ and $(f, 1) = \overline{\omega(f)}$, we see that

$$\frac{\partial^2 P_\Lambda}{\partial \lambda_X^2} = \mathbb{E}\left[J_X^2[(\Phi_X - \omega(\Phi_X), \Phi_X - \omega(\Phi_X))]\right] \geq 0. \tag{2.62}$$

This yields the following result.

Theorem 2.13 (Type I inequality, centered general quantum case) *For systems described by the quantum Hamiltonian (2.55) with general centered random interactions $\{J_X\}$ with $\mathbb{E}[J_X] = 0$, the following inequality holds: for all $X \subset \Lambda$ and for $\lambda_X \geqslant 0$*

$$\frac{\partial P_\Lambda}{\partial \lambda_X} = \langle J_X \Phi_X \rangle \geqslant 0. \tag{2.63}$$

Proof The argument we use goes essentially along the same lines as the classical case. The only additional technicality is represented by the Duhamel inner product, a basic tool for rigorous results in quantum statistical mechanics. We first observe that since the second derivative of the pressure is non-negative,

$$\frac{\partial^2 P_\Lambda}{\partial \lambda_X^2} \geq 0. \tag{2.64}$$

We deduce that the first derivative

$$\frac{\partial P_\Lambda}{\partial \lambda_X} = \mathbb{E}[J_X \omega(\Phi_X)] \tag{2.65}$$

is a monotonic non-decreasing function of λ_X. As a consequence, we have that for $\lambda_X \geqslant 0$

$$\frac{\partial P_\Lambda}{\partial \lambda_X} \geqslant \mathbb{E}[J_X \omega(\Phi_X)]\,|_{\lambda_X=0}, \tag{2.66}$$

but for $\lambda_X = 0$ the two random variables J_X and $\omega(\Phi_X)$ are independent:

$$\mathbb{E}[J_X \omega(\Phi_X)]\,|_{\lambda_X=0} = \mathbb{E}[J_X] \cdot \mathbb{E}[\omega(\Phi_X)]\,|_{\lambda_X=0} = 0 \tag{2.67}$$

where the last equality comes from having chosen distributions with $\mathbb{E}[J_X] = 0$. It also follows that for $\lambda_X \leqslant 0$ one has $\mathbb{E}[J_X \omega(\Phi_X)] \leqslant 0$. □

2.5 Type II inequalities: results and counterexamples

In analogy with the ferromagnetic case, in which type II inequalities are used to further investigate the properties of the correlation function, the monotonic behavior of the correlation functions has been investigated in the context of spin glasses. So far no general results have been found: the tested conjectures were all disproved.

Nevertheless, in the special case of one-dimensional systems (this Section) or in the spin glass models on the Nishimori line (Section 2.6), a type II inequality holds.

A natural problem is to look at the sign of the quantity

$$\frac{\partial^2 P_\Lambda}{\partial \lambda_X \partial \lambda_Y} = \mathbb{E}\left[J_X J_Y [\omega(\sigma_{XY}) - \omega(\sigma_X)\omega(\sigma_Y)]\right]. \tag{2.68}$$

By convexity, the previous quantity is obviously positive when $X = Y$. In Contucci and Unguendoli (2008) it has been proved that, for one-dimensional systems, the previous quantity is always negative for all $X \neq Y$. The precise statement also includes the extension of the sign of (2.68) to the positive values of the mean for the interaction variable. Consider a chain with periodic boundary condition

$$H_N(\sigma) = -\sum_{i=1}^{N} J_i \sigma_i \sigma_{i+1}$$

with $\sigma_{N+1} = \sigma_1$ with J_i independent one from each other. We assume they fulfill one of the three following hypotheses:

1. the random variables J_i are continuous with distributions $p(J_i)$ such that

$$p(|J_i|) \geq p(-|J_i|), \quad \forall i \text{ and } \forall |J_i| \in \mathbb{R}^+;$$

2. the J_i are symmetric around a positive mean $\mu_i > 0$:

$$p(\mu_i + |J_i|) = p(\mu_i - |J_i|), \quad \forall i \text{ and } \forall |J_i| \in \mathbb{R}^+;$$

3. the J_i are discrete variables taking on values $\pm J^{(i)}$ with $J^{(i)} > 0$ such that

$$p_i = p(J^{(i)}), \;\; q_i = p(-J^{(i)}),$$

and

$$\alpha := \prod_i (p_i - q_i) \geqslant 0.$$

In this section, ω_h denotes the thermal average of the quantity $\sigma_h \sigma_{h+1}$, and $\omega_{h,k}$ the thermal average of the quantity $\sigma_h \sigma_{h+1} \sigma_k \sigma_{k+1}$. The following theorem (see Contucci and Unguendoli (2008) for the proof) holds.

Theorem 2.14 (Type II inequality, one dimension) *For the one-dimensional system described above, one has:*

- *for all three hypotheses:*

$$\mathbb{E}[J_h \omega_h] > 0, \qquad \forall h = 1...N; \tag{2.69}$$

- *for hypotheses 1 and 3 with $\alpha = 0$:*

$$\mathbb{E}\,[J_h J_k(\omega_{hk} - \omega_h\omega_k)] < 0, \qquad \forall h, k = 1...N, \quad h \neq k; \tag{2.70}$$

- *for hypothesis 3, with $\alpha > 0$, the following properties hold:*
  $\forall l$, there exists in the $(J^{(l)}, \alpha)$ quadrant, a curve $\alpha(J^{(l)})$ such that the quantity

$$\mathbb{E}\,[J_h J_k(\omega_{hk} - \omega_h\omega_k)] \tag{2.71}$$

  changes its sign from negative to positive when crossing the curve $\alpha(J^{(l)})$ by increasing α, and such that on the curve $\alpha(J^{(l)})$

$$\mathbb{E}\,[J_h J_k(\omega_{hk} - \omega_h\omega_k)] = 0, \qquad \forall h, k = 1\ldots N, \quad h \neq k. \tag{2.72}$$

  Moreover $\mathbb{E}\,[J_h J_k(\omega_{hk} - \omega_h\omega_k)]$ is increasing in α along the $J^{(l)} = const.$ lines.

The result of the previous theorem for symmetric distribution does not hold for general lattice topologies. The quantity (2.68) has indeed been studied in Contucci *et al.* (2008) and a counterexample has been found for a six-spin system arranged in a one-dimensional topology with an extra interaction not belonging to the nearest-neighbor structure. What has been found is a change of sign from high to low temperatures.

Clearly the counterexample for a small system doesn't necessarily rule out the possibility of having a monotonic dependence of the correlation functions on the volume of the system (as happens with ferromagnets). To investigate that property in the same paper, a numerical simulation has been performed in two- and three-dimensional systems revealing that $\mathbb{E}\,[J_X\omega(\sigma_X)]$ is not monotonic in the volume.

Another interesting investigation, testing the analogy between magnetization on ferromagnetic models and overlaps in spin glasses, has been about the sign of the quenched expectation of the overlaps: namely, considering $q_X(\sigma, \tau) = \sigma_X\tau_X$, the sign of

$$\langle q_{XY}\rangle - \langle q_X\rangle\langle q_Y\rangle. \tag{2.73}$$

The topology used for the previous counterexample also provided the same phenomenon of temperature-dependent sign for this expression (Satoshi Morita, personal communication).

2.6 Inequalities of types I and II on the Nishimori line

We give here a definition of the Nishimori line that extends beyond the original one, given in Nishimori (1981, 2001), to the general Hamiltonian treated in this book.

Definition 2.15 (Nishimori line) Consider, for positive λ_X, the Hamiltonian

$$H_\Lambda(\sigma) = -\sum_{X\subset\Lambda} \lambda_X J_X \sigma_X \tag{2.74}$$

with mean value and variance of the interaction defined as

$$\mathbb{E}[J_X] = \mu_X, \tag{2.75}$$

$$\mathbb{E}\left[(J_X - \mu_X)^2\right] = \Delta_X^2. \tag{2.76}$$

If the distribution of the coupling is Gaussian, the *Nishimori line* is defined by a family of constraints among the distribution parameters of each J_X, and the strength of the interaction λ_X through the formula

$$\mu_X = \lambda_X \Delta_X^2. \tag{2.77}$$

A simple change of variable for Gaussian integrals ($\bar{J}_X = \lambda_X J_X$) shows that such a condition is equivalent to defining a spin glass model of Hamiltonian

$$H_\Lambda(\sigma) = -\sum_{X\subset\Lambda} \bar{J}_X \sigma_X, \tag{2.78}$$

where the new interaction variables are independent Gaussians, each having mean equal to variance. Introducing the positive parameter

$$\gamma_X = \mu_X^2/\Delta_X^2 \tag{2.79}$$

one has

$$\mathbb{E}\left[\bar{J}_X\right] = \gamma_X, \qquad \mathbb{E}\left[(\bar{J}_X - \gamma_X)^2\right] = \gamma_X \tag{2.80}$$

and the quenched pressure on the Nishimori line becomes

$$P_\Lambda(\gamma) = \mathbb{E}\left[\ln \sum_\sigma \exp^{\sum_X \bar{J}_X \sigma_X}\right], \tag{2.81}$$

where the functional dependence on the parameters γ's is entirely contained within the probability distribution.

Within this line there is a series of exact results that use the *gauge invariance* of the model (see Nishimori (1981, 2001)). The main observation is that the internal energy of the model can be exactly computed. The general argument works as in the following lemma (see Contucci *et al.* (2009b)).

Lemma 2.16 (Exact results on the Nishimori line) *On the Nishimori line $\mu_X = \lambda_X \Delta_X^2$ one has*

$$\mathbb{E}\left[(\Omega(J_X \sigma_X))\right] = \mu_X. \tag{2.82}$$

Proof By definition:

$$\mathbb{E}\left[\Omega(J_X\sigma_X)\right] = \int_{-\infty}^{\infty} \prod_Y \left(dJ_Y \frac{1}{\sqrt{2\pi}\Delta_Y} \exp\left(-\frac{(J_Y-\mu_Y)^2}{2\Delta_Y^2}\right)\right) \times \frac{\sum_\sigma J_X\sigma_X e^{\sum_Z \lambda_Z J_Z\sigma_Z}}{\sum_\sigma e^{\sum_Z \lambda_Z J_Z \sigma_Z}}. \tag{2.83}$$

We apply the gauge transformation

$$J_X \to J_X\tau_X, \quad \sigma_i \to \sigma_i\tau_i,$$

for all $i \in \Lambda$ and $X \subset \Lambda$, where τ_i is a "gauge" variable fixed to 1 or -1 at each $i \in X$ and $\tau_X = \prod_{i\in X}\tau_i$. This change of variable leaves the integral and sums in the above equation invariant. Then, only the J_Y in the exponent for the Gaussian weight changes:

$$\begin{aligned}
&\mathbb{E}\left[\Omega(J_X\sigma_X)\right] \\
&= \int_{-\infty}^{\infty} \prod_Y \left(dJ_Y \frac{1}{\sqrt{2\pi}\Delta_Y} \exp\left(-\frac{(J_Y\tau_Y-\mu_Y)^2}{2\Delta_Y^2}\right)\right) \cdot \frac{\sum_\sigma J_X\sigma_X e^{\sum_Z \lambda_Z J_Z\sigma_Z}}{\sum_\sigma e^{\sum_Z \lambda_Z J_Z \sigma_Z}} \\
&= \int_{-\infty}^{\infty} \prod_Y \left(dJ_Y \frac{1}{\sqrt{2\pi}\Delta_Y} \exp\left(-\frac{J_Y^2+\mu_Y^2}{2\Delta_Y^2}\right)\right) \exp\left(\sum_Y \frac{J_Y\mu_Y\tau_Y}{\Delta_Y^2}\right) \\
&\quad \times \frac{\sum_\sigma J_X\sigma_X e^{\sum_Z \lambda_Z J_Z\sigma_Z}}{\sum_\sigma e^{\sum_Z \lambda_Z J_Z \sigma_Z}}.
\end{aligned} \tag{2.84}$$

Since this expression holds for any assignment of ± 1 to τ_i, we may sum it up over all possible $\{\tau_i\}_i$ and divide the result by $2^{|\Lambda|}$,

$$\begin{aligned}
&\mathbb{E}\left[\Omega(J_X\sigma_X)\right] \\
&= \frac{1}{2^{|\Lambda|}} \int_{-\infty}^{\infty} \prod_Y \left(dJ_Y \frac{1}{\sqrt{2\pi}\Delta_Y} \exp\left(-\frac{J_Y^2+\mu_Y^2}{2\Delta_Y^2}\right)\right) \sum_\tau e^{\sum_Y J_Y\mu_Y\tau_Y/\Delta_Y^2} \\
&\quad \times \frac{\sum_\sigma J_X\sigma_X e^{\sum_Z \lambda_Z J_Z\sigma_Z}}{\sum_\sigma e^{\sum_Z \lambda_Z J_Z \sigma_Z}}.
\end{aligned} \tag{2.85}$$

The sum over τ on the numerator and the sum over σ in the denominator cancel each other on the Nishimori line, resulting in the simplified expression

$$\begin{aligned}
&\mathbb{E}\left[\Omega(J_X\sigma_X)\right] \\
&= \frac{1}{2^{|\Lambda|}} \sum_\sigma \int_{-\infty}^{\infty} \prod_Y \left(dJ_Y \frac{1}{\sqrt{2\pi}\Delta_Y} \exp\left(-\frac{J_Y^2+\mu_Y^2}{2\Delta_Y^2}\right)\right) J_X\sigma_X e^{\sum_Z J_Z\sigma_Z\mu_Z/\Delta_Z^2}.
\end{aligned} \tag{2.86}$$

For given $\{\sigma_i\}_i$, let us change the integral variable according to $J_Y \to J_Y \sigma_Y$. Then σ disappears completely and the integral is just for the average of J_X

$$\begin{aligned}
&\mathbb{E}\left[\Omega(J_X \sigma_X)\right] \\
&= \frac{1}{2^{|\Lambda|}} \sum_{\sigma} \int_{-\infty}^{\infty} \prod_{Y} \left(dJ_Y \frac{1}{\sqrt{2\pi}\,\Delta_Y} \exp\left(-\frac{J_Y^2 + \mu_Y^2}{2\Delta_Y^2} \right) \right) J_X \, e^{\sum_Z J_Z \mu_Z / \Delta_Z^2} \\
&= \frac{1}{2^{|\Lambda|}} \cdot 2^{|\Lambda|} \cdot 1 \cdot \int_{-\infty}^{\infty} dJ_X \frac{1}{\sqrt{2\pi}\,\Delta_X} J_X \exp\left(-\frac{(J_X - \mu_X)^2}{2\Delta_X^2} \right) \\
&= \mathbb{E}\left[J_X\right].
\end{aligned} \tag{2.87}$$

□

Remark 2.17 The exact computation (2.82) on the Nishimori line can be shown to be equivalent to the identity

$$\mathbb{E}\left[\omega(\sigma_X)\right] = \mathbb{E}\left[\omega^2(\sigma_X)\right]. \tag{2.88}$$

This is an immediate consequence of integrations by parts for Gaussian variables with non-zero average

$$\mathbb{E}\left[J_X f(J)\right] = \mu_X \mathbb{E}\left[f(J)\right] + \Delta_X^2 \mathbb{E}\left[\frac{df(J)}{dJ_X}\right]. \tag{2.89}$$

Using the same method, one can prove (see Morita *et al.* (2004) and Morita *et al.* (2005)) other identities such as

$$\mathbb{E}\left[\omega(\sigma_X)\omega(\sigma_Y)\right] = \mathbb{E}\left[\omega(\sigma_X\sigma_Y)\omega(\sigma_Y)\right] = \mathbb{E}\left[\omega(\sigma_X)\omega(\sigma_Y)\omega(\sigma_X\sigma_Y)\right] \tag{2.90}$$

$$\mathbb{E}\left[\omega^2(\sigma_X)\omega(\sigma_Y)\right] = \mathbb{E}\left[\omega^2(\sigma_X)\omega^2(\sigma_Y)\right]. \tag{2.91}$$

A first approach one can follow to determine inequalities on the Nishimori line is to first compute the quantities (2.51) and (2.68), and then implement the Nishimori constraint. By (2.82)

$$\frac{\partial P_\Lambda}{\partial \lambda_X} = \mu_X \geqslant 0. \tag{2.92}$$

Nevertheless the cross derivative (2.68) does not have a sign that can be recognized by inspection. In fact, applying integration both by parts and by the identities (2.90), (2.91), one has

$$\frac{\partial^2 P_\Lambda}{\partial \lambda_X \partial \lambda_Y} = \left(\Delta_X^2 \delta_{X,Y} + \mu_X \mu_Y\right) \mathbb{E}\left[\omega(\sigma_X\sigma_Y) - \omega(\sigma_X)\omega(\sigma_Y)\right]; \tag{2.93}$$

where the sign of the truncated correlation function is unknown since the Griffiths–Kelly–Sherman II (2.34) inequalities do not apply to a model with alternating sign couplings. The attempts to extend them to the positive mean interactions case have so far been unsuccessful, apart from the result in one dimension.

A successful approach has been found instead by considering the expression of the pressure on the Nishimori line and computing derivatives with respect to the (only free) parameters γ_X in (2.81). Following Morita *et al.* (2004) and Morita *et al.* (2005) one has the following.

Theorem 2.18 (Inequalities on the Nishimori line)

$$\frac{\partial P_\Lambda}{\partial \gamma_X} = \frac{1}{2}(\mathbb{E}\left[\omega(\sigma_X)\right] + 1) \geq 0, \tag{2.94}$$

$$\frac{\partial^2 P_\Lambda}{\partial \gamma_X \partial \gamma_Y} = \frac{1}{2}\mathbb{E}\left[\left[\omega(\sigma_X\sigma_Y) - \omega(\sigma_X)\omega(\sigma_Y)\right]^2\right] \geq 0, \tag{2.95}$$

Proof The Gaussian distribution on the Nishimori line can be written, for each interaction J and parameter γ, as

$$G(\gamma_X) = \frac{1}{\sqrt{2\pi\gamma_X}} e - \frac{(\bar{J}_X - \gamma_X)^2}{2\gamma_X}. \tag{2.96}$$

A straightforward computation gives

$$G'(\gamma_X) = \left(-\frac{1}{2\gamma_X} + \frac{\bar{J}_X - \gamma_X}{\gamma_X} + \frac{(\bar{J}_X - \gamma_X)^2}{2\gamma_X^2}\right) G(\gamma_X). \tag{2.97}$$

Using the previous formula in the expression of the pressure we find

$$\frac{\partial P_\Lambda}{\partial \gamma_X} = -\frac{1}{2\gamma_X} P_\Lambda + \frac{1}{\gamma_X}\mathbb{E}\left[(\bar{J}_X - \gamma_X)\ln \mathcal{Z}_\Lambda\right] + \frac{1}{2\gamma_X^2}\mathbb{E}\left[(\bar{J}_X - \gamma_X)^2 \ln \mathcal{Z}_\Lambda\right], \tag{2.98}$$

and using integration by parts (2.89) we get

$$\frac{\partial P_\Lambda}{\partial \gamma_X} = \mathbb{E}\left[\omega(\sigma_X) + \frac{1}{2}\left(1 - \omega^2(\sigma_X)\right)\right]. \tag{2.99}$$

The first inequality (2.94) follows from the identity (2.88). For the second inequality we proceed in a similar way. We start from

$$\frac{\partial^2 P_\Lambda}{\partial \gamma_X \partial \gamma_Y} = \frac{1}{2}\frac{\partial \mathbb{E}\left[\omega(\sigma_X)\right]}{\partial \gamma_Y}. \tag{2.100}$$

Formula (2.97) and integration by parts (2.89) yield

$$\frac{1}{2}\frac{\partial \mathbb{E}\,[\omega(\sigma_X)]}{\partial \gamma_Y} = \frac{1}{2}\mathbb{E}\,[(\omega(\sigma_X\sigma_Y) - \omega(\sigma_X)\omega(\sigma_Y))(1-\omega(\sigma_Y))]\,. \quad (2.101)$$

Applying the identities (2.90) and (2.91), the second inequality (2.95) follows. □

2.7 Some consequences of correlation inequalities

The next chapter will deal with the use of the inequalities presented here to prove a set of results concerning the existence and monotonicity of the thermodynamic limit for the pressure per particle and also, on the Nishimori line, the existence of the correlation functions.

Here we state and prove some simple consequences which follow directly from the inequalities. They relate thermodynamic quantities on different lattices or dimensions.

Theorem 2.19 *Consider a spin glass model verifying the inequality (2.51). For a given volume Λ let $\mathcal{L}_\Lambda$ denote the set of subsets with non-zero interactions and*

$$H_\Lambda^{(\mathcal{L}_\Lambda)} = -\sum_{X\in\mathcal{L}_\Lambda} J_X\sigma_X\,. \quad (2.102)$$

If $\mathcal{L}_\Lambda \subset \mathcal{L}'_\Lambda$ then the pressure satisfies

$$P_\Lambda^{(\mathcal{L}_\Lambda)} \leqslant P_\Lambda^{(\mathcal{L}'_\Lambda)}\,. \quad (2.103)$$

Moreover, defining the d-dimensional cube of $\mathbb{Z}^d$ as $[1,L]^d$ one has

$$\frac{1}{L^d}P^{(d)}_{[1,L]^d} \geqslant \frac{1}{L^{(d-1)}}P^{(d-1)}_{[1,L]^{(d-1)}}\,. \quad (2.104)$$

Proof One can prove the result by interpolation with the parameters λ_X of (2.49). In particular, to show the first statement, introduce the interpolating Hamiltonian

$$H_\Lambda^{(\mathcal{L}'_\Lambda)}(t) = -\sum_{X\in\mathcal{L}'_\Lambda} \lambda_X J_X\sigma_X, \quad (2.105)$$

with

$$\lambda_X = \begin{cases} t, & \text{if } X\in\mathcal{L}'\backslash\mathcal{L},\\ 1, & \text{otherwise,}\end{cases} \quad (2.106)$$

and let $P_\Lambda(t)$ be the relative interpolating pressure. Thanks to inequality (2.51) we then obtain

$$P_\Lambda^{(\mathcal{L}'_\Lambda)} - P_\Lambda^{(\mathcal{L}_\Lambda)} = \int_0^1 \frac{dP_\Lambda(t)}{dt} dt = \sum_{X \in \mathcal{L}' \setminus \mathcal{L}} \int_0^1 \langle J_X \sigma_X \rangle dt \geqslant 0. \quad (2.107)$$

The second statement can be proved analogously by defining an interpolating Hamiltonian with parameter t on all the subsets that connect different $(d-1)$-dimensional layers of the d-dimensional cube. □

Example 2.20 From the previous theorem we deduce that the pressure of the Edwards–Anderson model (Definition 1.1, p. 3) with nearest-neighbor interaction is lower than that of the short-range model which also includes next-nearest-neighbor interactions.

Stronger results can be obtained on the Nishimori line, where a type II inequality was proved.

Theorem 2.21 *Consider a spin glass model on the Nishimori line (2.78). For a given volume Λ, let $\mathcal{L}_\Lambda$ denote the set of subsets with non-zero interactions and*

$$H_\Lambda^{(\mathcal{L}_\Lambda)} = - \sum_{X \in \mathcal{L}_\Lambda} J_X \sigma_X. \quad (2.108)$$

If $\mathcal{L}_\Lambda \subset \mathcal{L}'_\Lambda$ then the quenched pressure and the correlation functions satisfy

$$P_\Lambda^{(\mathcal{L}_\Lambda)} \leqslant P_\Lambda^{(\mathcal{L}'_\Lambda)}, \quad (2.109)$$

$$\frac{1}{L^d} P^{(d)}_{[1,L]^d} \geqslant \frac{1}{L^{(d-1)}} P^{(d-1)}_{[1,L]^{(d-1)}}, \quad (2.110)$$

$$\langle \sigma_X \rangle^{(\mathcal{L}_\Lambda)} \leqslant \langle \sigma_X \rangle^{(\mathcal{L}'_\Lambda)}, \quad (2.111)$$

$$\langle \sigma_X \rangle^{(d)} \geqslant \langle \sigma_X \rangle^{(d-1)}. \quad (2.112)$$

Proof Using the interpolation schemes of the previous theorem and the inequalities of Theorem 2.18. □

Example 2.22 The inequalities on correlations can be used for the argument about the location of the multicritical points, which are believed to lie on the Nishimori line for various lattices (see Nishimori (1981, 2001)). For example, let us consider three two-dimensional lattices: the triangular (TR), square (SQ), and hexagonal (HEX) lattices. The triangular lattice is obtained from the square lattice with the addition of bonds, and the square lattice is obtained from the hexagonal one. Thus, the magnetizations of the three lattices satisfy the following relation:

$$\langle \sigma_X \rangle_{\rm HEX} \le \langle \sigma_X \rangle_{\rm SQ} \le \langle \sigma_X \rangle_{\rm TR}. \tag{2.113}$$

Since the multicritical temperature is defined as

$$T_c = \sup\{T; \langle \sigma_i \rangle > 0\}, \tag{2.114}$$

we obtain

$$T_c^{\rm HEX} \le T_c^{\rm SQ} \le T_c^{\rm TR}. \tag{2.115}$$

Similarly, the multicritical temperature of the simple cubic lattice is higher than that of the square lattice because the former is constructed from the latter by adding interactions.

3

The infinite-volume limit

Abstract

In this chapter we deal with the large-volume limit, often called the thermodynamic limit. While the problem has received a lot of attention in deterministic systems from the early 1960s both in statistical mechanics (Ruelle (1999)) and Euclidean quantum field theory (Guerra (1972)), in random systems a major breakthrough has been the introduction of the quadratic interpolation method by Guerra and Toninelli (2002). We review the results for finite-dimensional systems with Gaussian or, more generally, centered interactions. We then extend the analysis to quantum models and to non-centered interactions satisfying a thermodynamic stability condition. Special attention is devoted to the correction to the leading term, i.e. the surface pressure, which is investigated for various boundary conditions and for a wide class of models. A complete result is obtained on the Nishimori line where we can make use of the full correlation inequalities set introduced in the previous chapter. Finally the mean-field case is analyzed for the relevant models that appear in the literature.

3.1 Introduction

The infinite-volume limit in spin glasses has been tackled for some time (Vuillermot (1977); Ledrappier (1977); Pastur and Figotin (1978); Khanin and Sinai (1979); van Enter and van Hemmen (1983); Zegarlinski (1991)). The difficulties with respect to the deterministic case arose due to the randomness of the interaction which requires both the study of the averaged quantities as well as the random ones. For quantities like pressure (free energy) and ground state energy per particle, the fluctuation between samples vanishes for large volumes (self-averaging), therefore it suffices to study the limit of the average value. The peculiarity of the spin glass models is that for other quantities, such as the overlap distribution, the fluctuations between samples do not vanish at large volumes (at least in mean-field theory), and interesting information is also obtained from the study of the random quantities.

While several approaches for the finite-dimensional case were found to solve specific cases, the mean-field case resisted until the pioneering work by Guerra and Toninelli (2002), where the interpolating technique was introduced and eventually led to the rigorous solution of the Sherrington–Kirkpatrick model (Guerra (2003); Talagrand (2006)). The approach we follow here is based on the inequalities we developed in the previous chapter and has the same flavor introduced by Griffiths for the ferromagnetic systems. It is based on (i) the identification of monotonicity properties for the thermodynamic functions, and (ii) their boundedness. This leads to the existence of a thermodynamic limit in the sense of Fisher (fixed shape growth). We show that both (i) and (ii) are structural properties of a large family of models and are based on convexity arguments and on *thermodynamic stability*. This last property in spin glasses emerges as a natural generalization of the equivalent property for deterministic systems (see Ruelle (1987)) and includes non-summable interactions extending the results by Khanin and Sinai (1979), and van Enter and van Hemmen (1983).

3.2 Finite-dimensional models with Gaussian interactions

Consider the case of a Gaussian spin glass defined by the Hamiltonian (1.8) with Gaussian-centered independent couplings:

$$\mathbb{E}\left[J_X\right] = 0, \tag{3.1}$$

and variance

$$\mathbb{E}\left[J_X^2\right] = \Delta_X^2. \tag{3.2}$$

From now on we shall always work with translation invariant couplings distributions, i.e. for all $a \in \mathbb{Z}^d$ and for all $X \subset \mathbb{Z}^d$

$$J_{T_a X} \stackrel{\mathcal{D}}{=} J_X \tag{3.3}$$

where $T_a X = \{i + a\}_{i \in X}$. Here we are following Contucci and Graffi (2004a).

Theorem 3.1 (Thermodynamic limit, Gaussian case) *If the condition of* thermodynamic stability *is fulfilled*

$$\sup_{\Lambda} \frac{1}{|\Lambda|} \sum_{X \subset \Lambda} \Delta_X^2 \leqslant \bar{c} < +\infty \tag{3.4}$$

then the following limit exists and is finite

$$\lim_{\Lambda \nearrow \mathbb{Z}^d} \frac{P_\Lambda}{|\Lambda|} = \sup_\Lambda \frac{P_\Lambda}{|\Lambda|} = p, \tag{3.5}$$

where the limit is in the sense of Fisher.

Proof The constituents of the proof are the thermodynamic stability hypothesis (3.4) and the lemmata that we prove after the theorem.

For simplicity we will consider growing cubes. Their existence in the sense of Fisher (see Ruelle (1987); Fisher (1964) for more general volumes, parallelepiped, etc.) can be easily deduced. Let $\Lambda = [1, N]^d$ be the d-dimensional cube of side N. For a given $M < N$ let m and r be divisor and rest as in $N = mM + R$, where the integer $m \geqslant 1$ and $0 \leqslant r \leqslant M - 1$. Specializing the superadditivity property of Lemma 3.2 to the case of cubic lattices with translation invariant couplings distribution, we have

$$P_{[1,mM]^d} \geqslant m^d P_{[1,M]^d}. \tag{3.6}$$

Moreover, from Lemma 3.3, we have

$$P_{[1,N]^d} \geqslant m^d P_{[1,M]^d} + (N^d - (mM)^d) \ln 2. \tag{3.7}$$

Now we let $N \to \infty$ for M fixed (i.e. $m \to \infty$). Since r is bounded by M, then $mM/N \to 1$ and $r/mM \to 0$. By consequence

$$\liminf_{N\to\infty} \frac{P_{[1,N]^d}}{N^d} \geqslant \frac{P_{[1,M]^d}}{M^d}. \tag{3.8}$$

Since the previous inequality is true for all M it implies

$$\liminf_{N\to\infty} \frac{P_{[1,N]^d}}{N^d} \geqslant \sup_M \frac{P_{[1,M]^d}}{M^d} \geqslant \limsup_{N\to\infty} \frac{P_{[1,N]^d}}{N^d}, \tag{3.9}$$

i.e. the limit exists and is equal to the supremum $\sup_M \frac{P_{[1,M]^d}}{M^d}$ which, of course, could be infinite. To exclude this case we use Lemma 3.4. □

Lemma 3.2 (Pressure superadditivity) *Consider a partition of the set Λ into n non-empty disjoint sets Λ_s:*

$$\Lambda = \bigcup_{s=1}^{n} \Lambda_s, \tag{3.10}$$

$$\Lambda_s \cap \Lambda_{s'} = \emptyset. \tag{3.11}$$

The quenched pressure is superadditive:

$$P_\Lambda \geqslant \sum_{s=1}^{n} P_{\Lambda_s}. \tag{3.12}$$

Proof For each partition, the potential generated by all interactions among different subsets is defined as

$$\tilde{H}_\Lambda = H_\Lambda - \sum_{s=1}^{n} H_{\Lambda_s}; \tag{3.13}$$

from Eq. (1.8) we have that

$$\tilde{H}_\Lambda = \sum_{X \in \mathcal{C}_\Lambda} J_X \sigma_X \tag{3.14}$$

where the "corridors" $\mathcal{C}_\Lambda$ are the set of all $X \subset \Lambda$ which are not proper subsets of any Λ_s.

To each partition of Λ we associate the interpolating potential for $0 \leqslant t \leqslant 1$

$$H_\Lambda(t) = \sum_{s=0}^{n} t_s \, H_{\Lambda_s}, \tag{3.15}$$

with $t_0 = t$, $t_s = (1-t)$ for $1 \leqslant s \leqslant n$, $H_{\Lambda_0} = H_\Lambda$ and

$$H_{\Lambda_s} = \sum_{X \subset \Lambda_s} J_X \sigma_X. \tag{3.16}$$

We define the interpolating random partition function

$$\mathcal{Z}_\Lambda(t) = \sum_\sigma e^{-H_\Lambda(t)}, \tag{3.17}$$

and we observe that

$$\mathcal{Z}_\Lambda(0) = \prod_{s=1}^{n} \mathcal{Z}_{\Lambda_s}, \quad \mathcal{Z}_\Lambda(1) = \mathcal{Z}_\Lambda. \tag{3.18}$$

Consider the interpolating pressure

$$P_\Lambda(t) := \mathbb{E}\left[\ln \mathcal{Z}_\Lambda(t)\right] \tag{3.19}$$

and the corresponding states $\omega_t(-)$ and $\langle - \rangle_t$ (as in Definition 1.7, p. 6 with Hamiltonian $H_\Lambda(t)$). Thanks to (3.18) we get

$$P_\Lambda(0) = \sum_{s=1}^{n} P_{\Lambda_s}, \quad P_\Lambda(1) = P_\Lambda. \tag{3.20}$$

We observe now that

$$\frac{d}{dt} P_\Lambda(t) = \sum_{s=0}^{n} \epsilon_s \langle H_{\Lambda_s} \rangle_t, \tag{3.21}$$

with $\epsilon_0 = 1$ and $\epsilon_s = -1$ for $1 \leqslant s \leqslant n$. For each s we have

$$\langle H_{\Lambda_s} \rangle_t = \sum_{X \subset \Lambda_s} \langle J_X \sigma_X \rangle_t; \tag{3.22}$$

Summing up all the contributions in (3.21):

$$\frac{d}{dt} P_\Lambda(t) = \langle \tilde{H}_\Lambda \rangle_t = \sum_{X \in C_\Lambda} \langle J_X \sigma_X \rangle_t \geqslant 0, \tag{3.23}$$

thanks to the inequality (2.45) of Theorem 2.10 for the model of Hamiltonian $H_\Lambda(t)$. From (3.20) and (3.23), applying the fundamental theorem of calculus, we immediately get formula (3.12). □

Lemma 3.3 (Pressure lower bound) *For any $\Lambda \subset \mathbb{Z}^d$ one has the bound*

$$P_\Lambda \geqslant |\Lambda| \ln 2. \tag{3.24}$$

Proof The strategy is similar to the previous lemma; one now uses an interpolation of the form $\{t J_X\}$ for all $X \in \Lambda$ for $t \in [0, 1]$. The inequality in Theorem 2.10 implies the result. □

Lemma 3.4 (Pressure upper bound) *For any $\Lambda \subset \mathbb{Z}^d$ one has the bound*

$$P_\Lambda \leqslant |\Lambda| \left(\ln 2 + \frac{\bar{c}}{2} \right). \tag{3.25}$$

Proof Using Jensen's inequality

$$P_\Lambda = \mathbb{E}[\ln \mathcal{Z}_\Lambda] \leqslant \ln \mathbb{E}[\mathcal{Z}_\Lambda] = \ln \sum_\sigma \mathbb{E}\left[e^{-H_\Lambda}\right]. \tag{3.26}$$

Since for a centered Gaussian random variable one has

$$\mathbb{E}\left[e^{-H_\Lambda}\right] = e^{\frac{1}{2}\mathbb{E}[H_\Lambda^2]} \tag{3.27}$$

and from the hypothesis of thermodynamic stability (3.4)

$$\frac{\mathbb{E}\left[H_\Lambda^2\right]}{2} \leqslant \frac{1}{2} \sum_{X \subset \Lambda} \Delta_X^2 \leqslant \frac{\bar{c}}{2} |\Lambda|, \tag{3.28}$$

we obtain the lemma. □

Let us verify the stability condition (3.4) in the following examples.

Example 3.5 Edwards–Anderson model (Definition 1.1). For a volume $\Lambda \subset \mathbb{Z}^d$

$$\sum_{X \subset \Lambda} \Delta_X^2 = \sum_{||i-j||=1} \Delta_{(i,j)}^2 \leqslant d|\Lambda| \tag{3.29}$$

since $\Delta_{(i,j)} = 1$. The equality is reached if Λ is chosen to be the cubic box and periodic boundary conditions are imposed.

Example 3.6 More generally, for a volume $\Lambda \subset \mathbb{Z}^d$ and the case of power law interactions:

$$J_{i,j} = \frac{\tilde{J}_{i,j}}{||i-j||^{d\alpha}} \tag{3.30}$$

where $\tilde{J}_{i,j}$ are i.i.d. standard Gaussian, one has

$$\sum_{X\subset\Lambda} \Delta_X^2 = \sum_{(i,j)\in\Lambda\times\Lambda} \frac{1}{||i-j||^{2d\alpha}} \leqslant c_d|\Lambda| \tag{3.31}$$

with

$$c_d = \int_1^\infty dx_1 \cdots \int_1^\infty dx_d \frac{1}{\left(x_1^2 + \cdots + x_d^2\right)^{d\alpha}}. \tag{3.32}$$

Therefore the stability condition (3.4) is satisfied if $\alpha > 1/2$, thus including square summable interactions.

Remark 3.7 It is interesting to observe that the interpolating strategy also applies to standard ferromagnetic systems. Consider for instance the ferromagnetic model ($J_X \geqslant 0$) with Hamiltonian (2.25); the proof of the existence of the thermodynamic limit of the pressure would basically read the same up to two observations:

1. the model with interaction tJ_X with positive t is still ferromagnetic so that $\omega_t(\sigma_X) \geqslant 0$, due to the first Griffiths inequality;
2. the condition of thermodynamic stability is replaced by the standard one

$$\sup_\Lambda \frac{1}{|\Lambda|} \sum_{X\subset\Lambda} J_X \leqslant \bar{c} < +\infty. \tag{3.33}$$

A similar condition of *simple summability*, as opposed to the previous one (3.4) of square summability, will emerge again for the non-centered distribution of interactions in Section 3.6.

Definition 3.8 (Ground state energy) The random ground state energy of the model with Hamiltonian H_Λ is defined as

$$\mathcal{E}_\Lambda = \inf_{\sigma\in\Sigma_\Lambda} H_\Lambda(\sigma), \tag{3.34}$$

and its mean value is given by

$$E_\Lambda = \mathbb{E}\left[\mathcal{E}_\Lambda\right]. \tag{3.35}$$

The properties of E_Λ can be related to those of P_Λ by rescaling the Hamiltonian with the positive factor (inverse temperature) β, as in thermodynamics. By general thermodynamic arguments (positivity of the entropy) one has

$$\lim_{\beta\to\infty} -\frac{P_\Lambda(\beta)}{\beta} \searrow E_\Lambda. \tag{3.36}$$

Therefore one obtains

$$E_\Lambda \leqslant E_{\Lambda_1} + E_{\Lambda_2}, \tag{3.37}$$

which implies the following.

Corollary 3.9

$$e = \lim_{\Lambda\nearrow\mathbb{Z}^d} \frac{E_\Lambda}{|\Lambda|} = \inf_\Lambda \frac{E_\Lambda}{|\Lambda|}, \tag{3.38}$$

and

$$\lim_{\beta\to\infty} -\frac{p(\beta)}{\beta} \searrow e. \tag{3.39}$$

where $p(\beta)$ is the infinite-volume pressure (see Eq. (3.5)).

3.3 Pressure self-averaging

Having established the existence of a thermodynamic limit of the quenched free energy, a further question concerns the random (sample dependent) free energy. The result that one expects on a physical base is the vanishing of fluctuations between samples in the thermodynamic limit, often called the *self-averaging* property. This can be obtained either by martingale arguments (see Pastur and Shcherbina (1991) for the Sherrington–Kirkpatrick model, and Contucci and Giardinà (2007) for finite-dimensional models) or by concentration of measure (as in Talagrand (2010b), and Guerra and Toninelli (2003)). Here we follow the second approach. Our formulation applies to both mean-field and finite-dimensional models and includes, for instance, the non-summable interactions in finite dimensions and the p-spin mean-field models, as well as the random energy models (REM) and generalized random energy models (GREM).

Theorem 3.10 (Pressure self-averaging) *Consider the Gaussian spin glass defined by the Hamiltonian (1.8) with Gaussian-centered independent couplings of variance $\mathbb{E}\left[J_X^2\right] = \Delta_X^2$. If the condition of* thermodynamic stability *(3.4) is fulfilled, then the disorder fluctuation of the pressure satisfies the following inequality: for* *all* $x > 0$

$$\mathbb{P}(|\mathcal{P}_\Lambda - \mathbb{E}[\mathcal{P}_\Lambda]| \geqslant x) \leqslant 2\exp\left(-\frac{x^2}{2\beta^2\bar{c}|\Lambda|}\right). \tag{3.40}$$

In particular

$$V(\mathcal{P}_\Lambda) = \mathbb{E}\left[\mathcal{P}_\Lambda^2\right] - \mathbb{E}[\mathcal{P}_\Lambda]^2 \leqslant 4\bar{c}\beta^2\,|\Lambda|. \tag{3.41}$$

Remark 3.11 From the concentration inequality (3.40), the random free energy almost surely converges to its limit too, as it follows from an immediate application of the Borel–Cantelli lemma.

Proof Consider an $s > 0$. By Markov's inequality, one has

$$\begin{aligned}\mathbb{P}\,\{\mathcal{P}_\Lambda - \mathbb{E}[\mathcal{P}_\Lambda] \geqslant x\} &= \mathbb{P}\,\{\exp[s(\mathcal{P}_\Lambda - \mathbb{E}[\mathcal{P}_\Lambda])] \geqslant \exp(sx)\} \\ &\leqslant \mathbb{E}\left[\exp[s(\mathcal{P}_\Lambda - \mathbb{E}[\mathcal{P}_\Lambda])]\right]\exp(-sx).\end{aligned} \tag{3.42}$$

To bound the generating function

$$\mathbb{E}\left[\exp[s(\mathcal{P}_\Lambda - \mathbb{E}[\mathcal{P}_\Lambda])]\right] \tag{3.43}$$

one introduces, for a parameter $t \in [0, 1]$, the following interpolating function:

$$\phi(t) = \ln \mathbb{E}_1\{\exp(s\ \mathbb{E}_2\{\ln \mathcal{Z}(t)\})\}, \tag{3.44}$$

where $\mathbb{E}_1\{-\}$ and $\mathbb{E}_2\{-\}$ denote expectation with respect to two independent copies $X_1(\sigma)$ and $X_2(\sigma)$ of the Hamiltonian $H_\Lambda(\sigma)$, and the partition function $Z(t)$ is

$$\mathcal{Z}(t) = \sum_{\sigma\in\Sigma_\Lambda} e^{-\beta\sqrt{t}X_1(\sigma)-\beta\sqrt{1-t}X_2(\sigma)}. \tag{3.45}$$

It can immediately be verified that

$$\phi(0) = s\mathbb{E}[\mathcal{P}_\Lambda], \tag{3.46}$$

and

$$\phi(1) = \ln \mathbb{E}\left[e^{s\,\mathcal{P}_\Lambda}\right]. \tag{3.47}$$

This implies that

$$\mathbb{E}\left[\exp[s(\mathcal{P}_\Lambda - \mathbb{E}[\mathcal{P}_\Lambda])]\right] = e^{\phi(1)-\phi(0)} = e^{\int_0^1 \phi'(t)dt}. \tag{3.48}$$

On the other hand, the derivative with respect to t can be easily bounded. Defining

$$K(t) = \exp(s\ \mathbb{E}_2\{\ln \mathcal{Z}(t)\}) \tag{3.49}$$

and

$$p^t(\sigma) = \frac{e^{-\beta\sqrt{t}X_1(\sigma)-\beta\sqrt{1-t}X_2(\sigma)}}{\mathcal{Z}(t)} \tag{3.50}$$

one has

$$\phi'(t) = \frac{\mathbb{E}_1\Big\{K(t)\; s\beta\; \mathbb{E}_2\Big\{\sum_\sigma p^t(\sigma)\Big[\frac{1}{2\sqrt{t}}X_1(\sigma) - \frac{1}{2\sqrt{1-t}}X_2(\sigma)\Big]\Big\}\Big\}}{\mathbb{E}_1\{K(t)\}}. \tag{3.51}$$

Applying the integration by parts formula (1.19), a simple computation gives

$$\begin{aligned}
&\sum_\sigma \mathbb{E}_1\left\{K(t)\,\mathbb{E}_2\left\{p^t(\sigma)\frac{1}{\sqrt{t}}\,X_1(\sigma)\right\}\right\} \\
&\quad = s\beta\sum_{\sigma,\tau}\mathbb{E}_1\left\{K(t)\,\mathcal{C}_\Lambda(\sigma,\tau)\,\mathbb{E}_2\left\{p^t(\tau)\right\}\,\mathbb{E}_2\left\{p^t(\sigma)\right\}\right\} \\
&\qquad + \beta\mathbb{E}_1\left\{K(t)\,\mathbb{E}_2\left\{\sum_\sigma \mathcal{C}_\Lambda(\sigma,\sigma)p^t(\sigma)\right\}\right\} \\
&\qquad - \beta\mathbb{E}_1\left\{K(t)\,\mathbb{E}_2\left\{\sum_{\sigma,\tau}\mathcal{C}_\Lambda(\sigma,\tau)p^t(\sigma)p^t(\tau))\right\}\right\}
\end{aligned}$$

and

$$\begin{aligned}
&\mathbb{E}_1\left\{K(t)\,\mathbb{E}_2\left\{\sum_\sigma p^t(\sigma)\frac{1}{\sqrt{1-t}}\,X_2(\sigma)\right\}\right\} \\
&\quad = \beta\mathbb{E}_1\left\{K(t)\,\mathbb{E}_2\left\{\sum_\sigma \mathcal{C}_\Lambda(\sigma,\sigma)p^t(\sigma)\right\}\right\} \\
&\qquad - \beta\mathbb{E}_1\left\{K(t)\,\mathbb{E}_2\left\{\sum_{\sigma,\tau}\mathcal{C}_\Lambda(\sigma,\tau)p^t(\sigma)p^t(\tau))\right\}\right\}.
\end{aligned}$$

Taking the difference between the previous two expressions, one finds

$$\phi'(t) = \frac{s^2\beta^2}{2}\frac{\sum_{\sigma,\tau}\mathbb{E}_1\left\{K(t)\,\mathcal{C}_\Lambda(\sigma,\tau)\,\mathbb{E}_2\left\{p^t(\tau)\right\}\,\mathbb{E}_2\left\{p^t(\sigma)\right\}\right\}}{\mathbb{E}_1\{K(t)\}}. \tag{3.52}$$

Using the thermodynamic stability condition (3.4), this yields

$$|\phi'(t)| \leqslant \frac{s^2\beta^2}{2}|\Lambda|\bar{c} \tag{3.53}$$

from which it follows

$$\mathbb{E}\left[\exp[s(\mathcal{P}_\Lambda - \mathbb{E}[\mathcal{P}_\Lambda])]\right] \leqslant \exp\left(\frac{s^2\beta^2}{2}|\Lambda|\bar{c}\right). \tag{3.54}$$

Inserting this bound into the inequality (3.42) and optimizing over s, one finally obtains

$$\mathbb{P}\left(\mathcal{P}_\Lambda - \mathbb{E}[\mathcal{P}_\Lambda] \geqslant x\right) \leqslant \exp\left(-\frac{x^2}{2\bar{c}\beta^2|\Lambda|}\right). \tag{3.55}$$

The proof of inequality (3.40) is completed by observing that one can repeat a similar computation for $\mathbb{P}\left(\mathcal{P}_\Lambda - \mathbb{E}[\mathcal{P}_\Lambda] \leqslant -x\right)$. The result for the variance (3.41) is then immediately proved using the identity

$$\mathbb{E}\left[(\mathcal{P}_\Lambda - \mathbb{E}[\mathcal{P}_\Lambda])^2\right] = 2\int_0^\infty x\, \mathbb{P}(|\mathcal{P}_\Lambda - \mathbb{E}[\mathcal{P}_\Lambda]| \geqslant x)\, dx. \tag{3.56}$$

□

Corollary 3.12 (Ground state self-averaging) *The disorder fluctuation of the ground state energy (see Definition 3.8) satisfies the following inequality: for all* $x > 0$

$$\mathbb{P}\left(|\mathcal{E}_\Lambda - \mathbb{E}[\mathcal{E}_\Lambda]| \geqslant x\right) \leqslant 2\exp\left(-\frac{x^2}{2\bar{c}|\Lambda|}\right). \tag{3.57}$$

In particular the ground state energy is self-averaging, i.e.

$$\mathrm{Var}(\mathcal{E}_\Lambda) = \mathbb{E}\left[\mathcal{E}_\Lambda^2\right] - \mathbb{E}[\mathcal{E}_\Lambda]^2 \leqslant 4\,\bar{c}\,|\Lambda|. \tag{3.58}$$

Proof The proof derives simply from Eq. (3.36). □

3.4 Finite-dimensional models with centered interactions

The case of general random interaction J_X in the Hamiltonian (1.8) can be treated in analogy with the Gaussian case, but for the condition of thermodynamic stability which we will implement in different forms. For the superadditivity of the pressure, we notice that it works with no difference with respect to the Gaussian case.

Lemma 3.13 (Pressure superadditivity, general centered case) *Consider a partition of* Λ *into* n *non-empty disjoint sets* Λ_s. *For the general model (1.8) with centered interactions* $\mathbb{E}[J_X] = 0$, *the quenched pressure is superadditive:*

$$P_\Lambda \geqslant \sum_{s=1}^{n} P_{\Lambda_s}. \tag{3.59}$$

Proof This is identical to the Gaussian case in Lemma 3.2, and is guaranteed by the inequality (2.51) for general centered interaction. □

Remark 3.14 As far as the superadditivity property is concerned, the zero average hypothesis can be relaxed on the sets X with $|X| = 1$. In fact, the proof only uses the positivity of the gauge invariant correlation function along "corridors" which only include contributions from subsets X with $|X| \geqslant 2$.

In order to obtain the boundedness of the pressure, the following is a possible condition.

Lemma 3.15 (Pressure upper bound, centered general case, exponential thermodynamic stability) *Consider the general model (1.8) with centered interactions* $\mathbb{E}[J_X] = 0$. *If the model fulfills the condition of exponential thermodynamic stability*

$$\mathbb{E}\left[e^{-H_\Lambda}\right] \leqslant e^{|\Lambda|\bar{c}} \tag{3.60}$$

with $\bar{c} < \infty$, *then the quenched pressure admits the bound*

$$P_\Lambda \leqslant |\Lambda| (\ln 2 + \bar{c}) \,. \tag{3.61}$$

Proof From Jensen's inequality and condition (3.60). □

For the Gaussian case, the previous condition is equivalent to the square summability of the interactions and is based on the bound for the *annealed* pressure $\ln \mathbb{E}[Z]$. We can, however, reach more general results in which the annealed pressure is unbounded but the quenched one does still have a thermodynamic limit. Here we use Contucci and Starr (2009), starting with the following result.

Lemma 3.16 (Bounds by recursion) *For the general model (1.8) one has*

$$P_\Lambda(\beta) \leq |\Lambda| \ln(2) + \sum_{X \subseteq \Lambda} (\mathbb{E}[\ln \cosh(J_X)] + |\mathbb{E}[\tanh(J_X)]|) \,. \tag{3.62}$$

Proof The proof is based on the simple observation which relates the pressure in the volume Λ to the pressure where the coupling of the subset X is set to zero. Introducing the tuning parameters $\{\lambda_X\}_{X \subset \Lambda}$ in the Hamiltonian

$$H_\Lambda(\{\lambda_X\}) = \sum_{X \subset \Lambda} \lambda_X J_X \sigma_X \tag{3.63}$$

one has

$$\frac{\mathcal{Z}_\Lambda}{\mathcal{Z}_\Lambda|_{\lambda_X = 0}} = \omega(e^{J_X \sigma_X})|_{\lambda_X = 0} \,. \tag{3.64}$$

Since by (2.13) one has

$$e^{J_X \sigma_X} = \cosh(J_X) + \sigma_X \sinh(J_X) \tag{3.65}$$

then (3.64) implies

$$\mathcal{P}_\Lambda - \mathcal{P}_\Lambda|_{\lambda_X=0} = \ln\cosh(J_X) + \ln(1 + \tanh(J_X)\cdot\omega(\sigma_X)|_{\lambda_X=0}) \tag{3.66}$$

Taking the average over disorder and using the inequality $1 + x \leq e^x$ and $\| \mathbb{E}\left[\omega(\sigma_X)|_{\lambda_X=0}\right] \| \leqslant 1$, it follows that

$$P_\Lambda - P_\Lambda|_{\lambda_X=0} \leqslant \mathbb{E}\left[\ln\cosh(J_X)\right] + |\mathbb{E}\left[\tanh(J_X)\right]|. \tag{3.67}$$

The theorem is obtained by iteration of the previous relation over all subsets $X \subset \Lambda$. □

Lemma 3.17 (Pressure upper bound, centered general case, square summable interactions) *Consider the general model (1.8) with centered interactions* $\mathbb{E}\left[J_X\right] = 0$. *If the couplings are square summable*

$$\|J\|_2^2 = \sum_{\substack{X\subset Z^d \\ X\ni 0}} \frac{\mathbb{E}[J_X^2]}{|X|} \leqslant \bar{c} < \infty, \tag{3.68}$$

then the quenched pressure admits the bound

$$P_\Lambda \leqslant |\Lambda| \left(\ln(2) + \frac{3}{2}\bar{c}\right). \tag{3.69}$$

Proof We observe that $\mathbb{E}[\ln\cosh(J_X)] \leq \mathbb{E}[J_X^2]/2$ because $\ln\cosh(x) \leq x^2/2$. Moreover, by assumption, $\mathbb{E}[J_X] = 0$. Therefore

$$|\mathbb{E}[\tanh(J_X)]| = |\mathbb{E}[\tanh(J_X) - J_X]| \leq \mathbb{E}\left[|\tanh(J_X) - J_X|\right].$$

But $|x - \tanh(x)| = \int_0^{|x|} \tanh^2(y)\,dy \leq |x|\tanh^2(x) \leq \min(|x|, |x|^3)$. It is easy to see that $\min(|x|, |x|^3) \leq x^2$. So we obtain $|\mathbb{E}[\tanh(J_X)]| \leq \mathbb{E}[J_X^2]$. Using the previous bounds in (3.62), together with assumption (3.68), gives the lemma. □

This corollary is comparable to the results of Khanin and Sinai (1979) and van Enter and van Hemmen (1983), except that we do not attempt to prove convergence in the van Hove sense, settling instead for convergence in the Fisher sense (see, for example, Ruelle (1999) for the difference). Also, we do not make any conditions on finite moments of the random couplings beyond existence of the variance. However, one can imagine a situation with fatter tails, so that even the variance does not exist. In that case, we can apply the following lemma.

Lemma 3.18 (Pressure upper bound, centered general case, summable interactions) *Consider the general model (1.8) with centered interactions $\mathbb{E}[J_X] = 0$. If the couplings are summable*

$$\|J\|_1 = \sum_{\substack{X \subset Z^d \\ X \ni 0}} \frac{\mathbb{E}[|J_X|]}{|X|} \leqslant \bar{c} < \infty, \tag{3.70}$$

then the quenched pressure admits the bound

$$P_\Lambda \leqslant |\Lambda| \left(\ln(2) + 2\bar{c}\right). \tag{3.71}$$

Proof Noting that $\ln\cosh(x) \le |x|$ and $|\tanh(x)| \le |x|$, we obtain from (3.62)

$$P_\Lambda \le |\Lambda| \left(\ln(2) + 2 \sum_{X \subseteq \Lambda} |J_X| \right),$$

which leads to the theorem by using (3.70). □

3.5 Quantum models

In this section we treat the quantum disordered systems introduced in Section 2.4. For convenience, we recall the definition of the quantum spin system general Hamiltonian

$$H_\Lambda = - \sum_{X \subset \Lambda} J_X \Phi_X, \tag{3.72}$$

where the operators Φ_X are self-adjoint elements of the real algebra generated by the set of *spin operators*, the Pauli matrices, $\sigma_i^{(x)}, \sigma_i^{(y)}, \sigma_i^{(z)}$, $i \in \Lambda$, on the Hilbert space $\mathcal{H}_X := \otimes_{i \in X} \mathcal{H}_i$, where $\mathcal{H}_i$ are copies of the finite-dimensional Hilbert space $\mathcal{H}$. The random interactions J_X are mutually independent with $\mathbb{E}[J_X] = 0$ for all X and $\mathbb{E}[J_X J_Y] = \Delta_X^2 \delta_{X,Y}$.

A first observation is that the quantum models obey to the same superadditivity property as the classical cases. Following Contucci and Lebowitz (2010) one can state the following.

Lemma 3.19 (Pressure superadditivity, quantum centered case) *Consider a partition of Λ into n non-empty disjoint sets Λ_s. For the general model (3.72) with centered interactions $\mathbb{E}[J_X] = 0$, the quenched pressure is superadditive:*

$$P_\Lambda \geqslant \sum_{s=1}^{n} P_{\Lambda_s}. \tag{3.73}$$

Proof One can proceed as in the Gaussian case of Lemma 3.2 and use the inequality (2.63) for general centered interactions. □

Remark 3.20 Following Contucci *et al.* (2004), it is interesting to observe that the superadditivity property that we have derived so far from the correlation inequality perspective can also be obtained directly from the convexity property of the pressure. This is true for all the disordered statistical mechanics models (classical and quantum) with centered interactions. In fact one can use the following lemma to cut the volume along the "corridors" (see Eq. (3.14)) and obtain (3.73):

Lemma 3.21 *Let $X_1, \dots X_n$ be independent random variables with zero mean. Let $F : \mathbb{R}^n \mapsto \mathbb{R}$ be such that for each $i = 1, \dots, n$ $x_i \mapsto F(x_1, \dots x_n)$ that is convex, then*

$$\mathbb{E}\left[F(X_1, \dots X_n)\right] \geq F(0, \dots 0) \tag{3.74}$$

where $\mathbb{E}$ denotes the expectation with respect to $X_1, \dots X_n$.

Proof This follows by applying Jensen's inequality to each X_i successively. □

For the the study of the boundedness – necessary to obtain the existence of the thermodynamic limit for the pressure – we follow Contucci *et al.* (2004); see also Zegarlinski (1991). Let

$$\|H\|_1 := \sum_{X \ni 0} \frac{\mathbb{E}\left[|J_X|\right] \|\Phi_X\|}{|X|} \tag{3.75}$$

and

$$\|H\|_2 := \left(\sum_{X \ni 0} \frac{\mathbb{E}\left[|J_X|^2\right] \|\Phi_X\|^2}{|X|} \right)^{1/2}, \tag{3.76}$$

where $\|\Phi_X\|$ denotes the operator norm.

Definition 3.22 We shall say that the random Hamiltonian H_Λ is stable if it is of the form

$$H_\Lambda = \tilde{H}_\Lambda + \hat{H}_\Lambda \tag{3.77}$$

where

$$\tilde{H}_\Lambda = -\sum_{X \subset \Lambda} \tilde{J}_X \tilde{\Phi}_X, \qquad \hat{H}_\Lambda = -\sum_{X \subset \Lambda} \hat{J}_X \hat{\Phi}_X$$

and all the $\tilde{J}_X$s and $\hat{J}_X$s are centered and independent, and $\|\tilde{H}\|_1$ and $\|\hat{H}\|_2$ are finite.

With this definition we shall prove in the next theorem that the specific pressure is bounded.

Lemma 3.23 *For the general model (3.72) with centered interactions $\mathbb{E}[J_X] = 0$ fulfilling the stability condition in Definition 3.22, the quenched specific pressure is bounded.*

Proof From the Bogoliubov inequality

$$\frac{\operatorname{Tr}(A-B)\,e^B}{\operatorname{Tr} e^B} \le \ln \operatorname{Tr} e^A - \ln \operatorname{Tr} e^B \le \frac{\operatorname{Tr}(A-B)\,e^A}{\operatorname{Tr} e^A} \tag{3.78}$$

with $A = H_\Lambda$ and $B = 0$ we get

$$\begin{aligned}
\log \mathcal{Z}_\Lambda - |\Lambda| \log \dim \mathcal{H} &\le -\frac{\operatorname{Tr} H_\Lambda\, e^{-H_\Lambda}}{\operatorname{Tr} e^{-H_\Lambda}} \\
&= -\frac{\operatorname{Tr} \tilde{H}_\Lambda\, e^{-H_\Lambda}}{\operatorname{Tr} e^{-H_\Lambda}} - \frac{\operatorname{Tr} \hat{H}_\Lambda\, e^{-H_\Lambda}}{\operatorname{Tr} e^{-H_\Lambda}} \\
&\le \|\tilde{H}_\Lambda\| - \frac{\operatorname{Tr} \hat{H}_\Lambda\, e^{-H_\Lambda}}{\operatorname{Tr} e^{-H_\Lambda}}.
\end{aligned} \tag{3.79}$$

Now

$$\mathbb{E}\left[\|\tilde{H}_\Lambda\|\right] \le |\Lambda| \|\tilde{H}\|_1. \tag{3.80}$$

For the other term we use the identity for the self-adjoint operators A and B

$$\frac{\operatorname{Tr} A\, e^{A+B}}{\operatorname{Tr} e^{A+B}} - \frac{\operatorname{Tr} A\, e^B}{\operatorname{Tr} e^B} = \int_0^1 dt\, (A - \langle A \rangle_t,\ A - \langle A \rangle_t)_t \tag{3.81}$$

where $\langle \cdot \rangle_t$ and $(\cdot,\cdot)_t$ denote the mean and the Duhamel inner product respectively with respect to $H = -(tA + B)$ in Eq. (2.61). The Duhamel inner product satisfies

$$(C, C) \le \frac{1}{2} \langle C^* C + C C^* \rangle^{1/2} \le \|C\|^2. \tag{3.82}$$

Therefore

$$\frac{\operatorname{Tr} A\, e^{A+B}}{\operatorname{Tr} e^{A+B}} - \frac{\operatorname{Tr} A\, e^B}{\operatorname{Tr} e^B} \le 4\|A\|^2. \tag{3.83}$$

With $A = \hat{J}_X \hat{\Phi}_X$ and $B = -H_\Lambda - \hat{J}_X \hat{\Phi}_X$ we get

$$\begin{aligned}
-\frac{\operatorname{Tr} \hat{H}_\Lambda\, e^{-H_\Lambda}}{\operatorname{Tr} e^{-H_\Lambda}} &= \sum_{X \subset \Lambda} \frac{\operatorname{Tr} \hat{J}_X \hat{\Phi}_X\, e^{-H_\Lambda}}{\operatorname{Tr} e^{-H_\Lambda}} \\
&\le \sum_{X \subset \Lambda} \operatorname{Tr} \hat{J}_X \hat{\Phi}_X \frac{e^{-H_\Lambda - \hat{J}_X \hat{\Phi}_X}}{\operatorname{Tr} e^{-H_\Lambda - \hat{J}_X \hat{\Phi}_X}} + 4 \sum_{X \subset \Lambda} |\hat{J}_X|^2 \|\hat{\Phi}_X\|^2.
\end{aligned} \tag{3.84}$$

Thus, since $-H_\Lambda - \hat{J}_X\hat{\Phi}_X$ is independent of $\hat{J}_X$ and $\mathbb{E}\left[\hat{J}_X\right] = 0$,

$$-\mathbb{E}\left[\frac{\operatorname{Tr}\hat{H}_\Lambda\, e^{-H_\Lambda}}{\operatorname{Tr} e^{-H_\Lambda}}\right] \le 4\sum_{X\subset\Lambda}\mathbb{E}\left[|\hat{J}_X|^2\right]\|\hat{\Phi}_X\|^2 \le 4|\Lambda|\|\hat{H}\|_2^2. \tag{3.85}$$

Therefore

$$P_\Lambda \le |\Lambda|\left(\log\dim\mathcal{H} + \|\tilde{H}\|_1 + 4\|\hat{H}\|_2^2\right). \tag{3.86}$$

□

3.6 Extension to non-centered interactions

In this section we show the existence of the thermodynamic limit of the pressure for general interactions which are allowed to have a non-zero average. For this case, a suitable hypothesis of summability on the interaction average values is required.

We treat the classical and quantum cases in parallel. Let $\|\Phi_X\|$ denote the supremum norm in the classical case and the operator norm in the quantum case. In the setting of general interactions, the pressure superadditivity property is replaced by the following one.

Lemma 3.24 (Superadditivity, general case) *Consider a partition of the set Λ into n non-empty disjoint sets Λ_s. For the model (3.72) with general interactions (i.e. possibly non-zero average), one has*

$$\bar{P}_\Lambda \geqslant \sum_{s=1}^{n}\bar{P}_{\Lambda_s}, \tag{3.87}$$

where

$$\bar{P}_\Lambda = P_\Lambda + \sum_{X\subset\Lambda,\,|X|>1}|\mathbb{E}\left[J_X\right]|\,\|\Phi_X\|. \tag{3.88}$$

Proof Let $\bar{J}_X := J_X - \mathbb{E}\left[J_X\right]$ for $|X| > 1$ so that $\bar{J}_X$ has zero mean and $\bar{J}_X := J_X$ if $|X| = 1$. Let

$$-H_\Lambda^{(1)} := \sum_{X\subset\Lambda}\bar{J}_X\Phi_X, \tag{3.89}$$

$$-H_\Lambda^{(2)} := \sum_{X\subset\Lambda,\,|X|>1}\left(\mathbb{E}\left[J_X\right]\Phi_X + |\mathbb{E}\left[J_X\right]|\,\|\Phi_X\|\right) \tag{3.90}$$

and

$$-\bar{H}_\Lambda := -H_\Lambda^{(1)} - H_\Lambda^{(2)}. \tag{3.91}$$

Then

$$-\bar{H}_\Lambda = -H_\Lambda + \sum_{X \subset \Lambda,\, |X|>1} |\mathbb{E}[J_X]|\, \|\Phi_X\|. \tag{3.92}$$

Let $\bar{P}_\Lambda$ be the pressure corresponding to $\bar{H}_\Lambda$. One can then see that $\bar{P}_\Lambda$ is superadditive by treating the terms in $H_\Lambda^{(1)}$ as in Lemma (3.13) in the classical case, or Lemma (3.19) in the quantum case, since each $\bar{J}_X$ has zero mean, except possibly if $|X| = 1$, and by using the fact that all the terms in $-H_\Lambda^{(2)}$ are positive. The superadditivity of $\bar{P}_\Lambda$ is then implied by the inequality

$$\operatorname{Tr} e^{(A+B)} \geq \operatorname{Tr} e^{A} \tag{3.93}$$

if B is a positive operator, which is elementary in the classical case and also holds in the quantum case (see, for instance, Simon (1993)). □

Lemma 3.24, combined with the boundedness of the specific pressure and a supplementary condition on the means, is sufficient to ensure the convergence of the specific pressure in the thermodynamic limit.

Theorem 3.25 (Thermodynamic limit, general case) *Consider the model (3.72) with general interactions (i.e. possibly non-zero average) fulfilling the stability condition in Definition 3.22. Then the thermodynamic limit of the quenched specific pressure pressure exists, is finite, and is equal to*

$$\lim_{\Lambda \nearrow \mathbb{Z}^d} \frac{P_\Lambda}{|\Lambda|} = \sup_{\Lambda} \frac{\bar{P}_\Lambda}{|\Lambda|} - c, \tag{3.94}$$

where

$$c := \sum_{X \ni 0,\, |X|>1} \frac{|\mathbb{E}[J_X]|\, \|\Phi_X\|}{|X|}. \tag{3.95}$$

The constant c is finite because of the stability assumption and the limit is in the sense of Fisher.

Proof The thermodynamic limit of $\bar{P}_\Lambda/|\Lambda|$ exists by superadditivity (Lemma 3.24) and boundedness. This last property follows from the stability assumption, together with Eq. (3.92) and the boundedness of P_Λ. The convergence of the specific pressure $P_\Lambda/|\Lambda|$ follows from (3.88) and, observing that definition (3.95) implies

$$\lim_{\Lambda \nearrow \mathbb{Z}^d} \frac{1}{|\Lambda|} \sum_{X \subset \Lambda,\, |X|>1} |\mathbb{E}[J_X]|\, \|\Phi_X\| = c. \tag{3.96}$$

□

3.7 Independence of the pressure from boundary conditions

Having established the existence of the pressure thermodynamic limit, one wonders about the effect of choosing different boundary conditions. A well-known property of finite-dimensional statistical mechanics models is the independence of the limiting pressure from boundary conditions. This is due to the vanishing of the ratio between the surface and volume of set Λ. There can be a problem with exceptional boundary conditions, as analysed by Gandolfi *et al.* (1993).

Definition 3.26 In a parallelepiped Λ with boundary $\partial\Lambda$, we consider for assigned boundary conditions Ξ the random partition function,

$$\mathcal{Z}_\Lambda^{(\Xi)} := \sum_\sigma e^{-H_\Lambda^{(\Xi)}}, \tag{3.97}$$

and the quenched pressure

$$P_\Lambda^{(\Xi)} := \mathbb{E}\left[\ln \mathcal{Z}_\Lambda^{(\Xi)}(J)\right]. \tag{3.98}$$

We will consider here three types of boundary conditions: free, periodic and antiperiodic. Other types (plus or minus boundary conditions, or overlap constraints) can be treated in a similar way.

1. In the free boundary conditions ϕ, the partition sum runs over all the spins inside the parallelepiped Λ:

$$\mathcal{Z}_\Lambda^{(\phi)} := \sum_\sigma e^{-H_\Lambda}. \tag{3.99}$$

2. In the periodic boundary conditions Π, the partition sum runs over all the spin values in the torus $\Pi_\Lambda = \mathbb{Z}^d/\Lambda$:

$$\mathcal{Z}_\Lambda^{(\Pi)} := \sum_\sigma e^{-H_{\Pi_\Lambda}}. \tag{3.100}$$

3. To define the antiperiodic conditions Π^*, we consider the Hamiltonian

$$H_{\Pi_\Lambda^*} = -\sum_{X\subset\Pi_\Lambda} J_X^*\sigma_X, \tag{3.101}$$

where, indicating with $\bar{\Lambda}$ the complementary set of Λ

$$J_X^* = J_X(-1)^{|X\cap\bar{\Lambda}|}. \tag{3.102}$$

Then, the partition function in the antiperiodic conditions Π^* is obtained by summing the Boltzmann–Gibbs measure associated to $H_{\Pi_\Lambda^*}$ over all the spin values in the torus $\Pi_\Lambda = \mathbb{Z}^d/\Lambda$:

$$\mathcal{Z}_\Lambda^{(\Pi_\Lambda^*)} := \sum_\sigma e^{-H_{\Pi_\Lambda^*}}. \tag{3.103}$$

The next theorem states the independence of the specific pressure from the choice of boundary conditions. To this end, it is necessary to assume a suitable condition on the coupling distribution. Here we will restrict our discussion to the Gaussian setting.

Theorem 3.27 (Pressure boundary independence) *Consider the Gaussian spin glass in a volume Λ defined by the Hamiltonian (1.8) with Gaussian-centered independent couplings of variance $\mathbb{E}\left[J_X^2\right] = \Delta_X^2$. Denoting Γ_Λ the set of all subsets of Λ, define the set*

$$\mathcal{C}_{\Pi_\Lambda} := \Gamma_{\Pi_\Lambda} \backslash \Gamma_\Lambda. \tag{3.104}$$

For $0 < \alpha < 1$, if the following condition holds

$$\sup_\Lambda \frac{1}{|\Lambda|^\alpha} \sum_{X \in \mathcal{C}_{\Pi_\Lambda}} \Delta_X^2 < \infty \tag{3.105}$$

then

$$\lim_{\Lambda \nearrow \mathbb{Z}^d} \frac{P_\Lambda^{(\phi)}}{|\Lambda|} = \lim_{\Lambda \nearrow \mathbb{Z}^d} \frac{P_\Lambda^{(\Pi)}}{|\Lambda|} = \lim_{\Lambda \nearrow \mathbb{Z}^d} \frac{P_\Lambda^{(\Pi^*)}}{|\Lambda|} = p. \tag{3.106}$$

Proof We first observe that, even for finite volume, the symmetry of the coupling distributions implies that $P_\Lambda^\Pi = P_\Lambda^{\Pi^*}$, from which the second equality of the theorem follows. To show the first equality, we introduce the interpolating pressure

$$P_\Lambda(t) = \mathbb{E}\left[\ln \sum_\sigma e^{-H_{\Pi_\Lambda}(t)}\right] \tag{3.107}$$

where

$$H_{\Pi_\Lambda}(t) = -\sum_{X \subset \Pi_\Lambda} \sqrt{t_X} J_X \sigma_X \tag{3.108}$$

with

$$t_X = \begin{cases} t, & \text{if } X \in \mathcal{C}_{\Pi_\Lambda}, \\ 1, & \text{otherwise.} \end{cases} \tag{3.109}$$

Observing that $P_\Lambda(1) = P_\Lambda^\Pi$ and $P_\Lambda(0) = P_\Lambda^\phi$ and applying integration by parts, one has

$$P_\Lambda^{(\Pi)} - P_\Lambda^{(\phi)} = \sum_{X \subset \mathcal{C}_{\Pi_\Lambda}} \Delta_X^2 \int_0^1 \left(1 - \langle q_X \rangle_t^{(\Pi_\Lambda)}\right) dt \tag{3.110}$$

where $q_X = \sigma_X \tau_X$ is the local overlap for the two spin configurations σ and τ, and $\langle - \rangle_t^{(\Pi_\Lambda)}$ is the quenched state corresponding to (3.108). Dividing by the volume $|\Lambda|$ and using the condition in Eq. (3.105), the first equality of the theorem follows. □

Example 3.28 In the d-dimensional Edwards–Anderson model (cf. Eq. (1.1)), one can check that the condition (3.105) is fulfilled for $\alpha = (d-1)/d$. In fact, for nearest-neighbor interactions $\mathcal{C}_{\Pi_\Lambda} = \partial\Lambda$ and (for cubes of size L) $|\partial\Lambda| = 2dL^{d-1}$; thus implying

$$\sum_{X \in \mathcal{C}_{\Pi_\Lambda}} \Delta_X^2 = 2dL^{d-1} \tag{3.111}$$

since $\Delta_X^2 = 1$.

Remark 3.29 The illustrated proof for Gaussian interactions can be extended to general interaction (3.72). In this case one can use the linear interpolation (as in Lemma 3.2) and obtain the result

$$P_\Lambda^{(\Pi)} - P_\Lambda^{(\phi)} = \sum_{X \subset \mathcal{C}_{\Pi_\Lambda}} \int_0^1 \langle J_X \omega(\sigma_X) \rangle_t^{(\Pi_\Lambda)} dt. \tag{3.112}$$

In such a case, the condition that guarantees independence from boundary conditions is still the same thanks to the bounds of Lemma 3.17. In case the interactions do not have a second moment, Lemma 3.18 still helps to prove the theorem by replacing the condition of square summability with simple summability.

3.8 Surface pressure

In statistical mechanics, once the existence of the thermodynamic limit has been taken care of for the free energy per unit volume, a natural next step is to establish at which rate – with respect to the volume – such a limit is reached. In particular, it is interesting to determine the next term in the expansion

$$\mathbb{E}[\ln \mathcal{Z}_\Lambda] = p|\Lambda| + o(|\Lambda|).$$

The problem has been analyzed since the pioneering work of Fisher and Lebowitz (1970) on classical particle systems, and followed by a series of results in both Euclidean quantum field theories (see Guerra (1972); Guerra *et al.* (1976)) and in ferromagnetic spin models (see Fisher and Caginalp (1977)). In those cases, the basic properties of monotonicity of the pressure and of correlations with respect to the strength of the interaction, namely the first and second Griffiths inequalities, made possible a rigorous proof of what thermodynamics suggests (see Simon

(1993)): for sufficiently regular potentials and (say) free boundary conditions, the pressure varies with the volume as a sum of a volume term plus a surface term called surface pressure. This surface term, unlike the volume one, depends in general not only on the interaction but also on the boundary conditions, and represents the contribution to the pressure due to the interaction of the system with its boundary.

We follow here Contucci and Graffi (2004b) and also Contucci and Lebowitz (2007, 2010).

Definition 3.30 In a parallelepiped Λ we define the quenched *surface pressure* with boundary condition Ξ as

$$T_\Lambda^{(\Xi)} := P_\Lambda^{(\Xi)} - p|\Lambda|, \tag{3.113}$$

where $P_\Lambda^{(\Xi)}$ is defined in (3.98) and p is the thermodynamic limit of the pressure per particle (in the sense of Fisher).

Let $k\Lambda$ be the k-magnified Λ – defined, for each positive integer k, as the parallelepiped in d dimensions of sides $kL_1, kL_2, \ldots, kL_d$. Consider the magnificated torus

$$\Pi_{k\Lambda} = \mathbb{Z}^d / k\Lambda. \tag{3.114}$$

For the tessellation $\{\Lambda_s\}$ of the torus $\Pi_{k\Lambda}$, where all the Λ_s are congruent to Λ, define the set

$$\mathcal{C}_{\Pi_{k\Lambda}} := \Gamma_{\Pi_{k\Lambda}} \setminus \bigcup_{s=1}^{k^d} \Gamma_{\Lambda_s}, \tag{3.115}$$

and associate with $\Pi_{k\Lambda}$ the interpolating potential

$$H_{\Pi_{k\Lambda}}(t) = - \sum_{X \subset \Pi_{k\Lambda}} \sqrt{t_X} J_X \sigma_X, \tag{3.116}$$

with

$$t_X = \begin{cases} t, & \text{if } X \in \mathcal{C}_{k\Pi_\Lambda}, \\ 1, & \text{otherwise.} \end{cases} \tag{3.117}$$

Finally, let $\langle - \rangle_t^{(\Pi_{k\Lambda})}$ be the corresponding quenched state.

Theorem 3.31 (Integral representation for $T_\Lambda^{(\phi)}$, Gaussian case) *The surface pressure admits the representation*

$$T_\Lambda^{(\phi)} = -\frac{1}{4} \lim_{k\to\infty} \sum_{X \subset \mathcal{C}_{\Pi_\Lambda}} \Delta_X^2 \int_0^1 (1 - \langle q_X \rangle_t^{(\Pi_{k\Lambda})}) dt. \tag{3.118}$$

In particular, if the following condition is fulfilled for some $0 < \alpha < 1$

$$\sup_{\Lambda} \frac{1}{|\Lambda|^{\alpha}} \sum_{X \in \mathcal{C}_{\Pi_\Lambda}} \Delta_X^2 \leqslant \bar{c} < \infty \tag{3.119}$$

then the quantity

$$\tau_\Lambda^{(\phi)} = \frac{T_\Lambda^{(\phi)}}{|\Lambda|^{\alpha}} \tag{3.120}$$

admits the bounds

$$-\frac{\bar{c}}{4} \leqslant \tau_\Lambda^{(\phi)} \leqslant 0. \tag{3.121}$$

Proof In finite volume and with free boundary conditions, we have by definition

$$P_\Lambda^{(\phi)} = \mathbb{E}\left[\ln \mathcal{Z}_\Lambda\right] = k^{-d}\mathbb{E}\left[\ln \mathcal{Z}_\Lambda^{k^d}\right]. \tag{3.122}$$

By Theorem 3.27, the limiting pressure per particle is independent of the boundary conditions. Hence:

$$p|\Lambda| = \lim_{k\to\infty} k^{-d}\mathbb{E}\left[\ln \mathcal{Z}_{k\Lambda}^{(\Pi)}\right]. \tag{3.123}$$

By (3.122) and (3.123) we obtain

$$T_\Lambda^{(\phi)} = \left(P_\Lambda^{(\phi)} - p|\Lambda|\right) = \lim_{k\to\infty} k^{-d}\mathbb{E}\left[\ln \mathcal{Z}_\Lambda^{k^d} - \ln \mathcal{Z}_{k\Lambda}^{(\Pi)}\right]. \tag{3.124}$$

For each $0 \leqslant t \leqslant 1$ we consider the interpolating Hamiltonian as in (3.116): the interpolating partition function

$$\mathcal{Z}^{(\Pi_{k\Lambda})}(t) = \sum_{\sigma} e^{-H^{(\Pi_{k\Lambda})}(t)}, \tag{3.125}$$

the interpolating pressure

$$P^{(\Pi_{k\Lambda})}(t) := \mathbb{E}\left[\ln \mathcal{Z}^{(\Pi_{k\Lambda})}(t)\right], \tag{3.126}$$

and the corresponding states $\omega_t^{(\Pi_{k\Lambda})}(-)$ and $\langle - \rangle_t^{(\Pi_{k\Lambda})}$. We observe that

$$\mathcal{Z}^{(\Pi_{k\Lambda})}(0) = \prod_{s=1}^{k^d} \mathcal{Z}_{\Lambda_s}, \quad \mathcal{Z}^{(\Pi_{k\Lambda})}(1) = \mathcal{Z}_{\Pi_{k\Lambda}}, \tag{3.127}$$

or equivalently

$$P^{(\Pi_{k\Lambda})}(0) = k^d P_\Lambda, \quad P^{(\Pi_{k\Lambda})}(1) = P_{\Pi_{k\Lambda}}, \tag{3.128}$$

and by (3.124)

$$T_\Lambda^{(\phi)} = \lim_{k\to\infty} k^{-d}\left[P^{(\Pi_{k\Lambda})}(0) - P^{(\Pi_{k\Lambda})}(1)\right] = -\lim_{k\to\infty} k^{-d}\int_0^1 \frac{d}{dt}P^{(\Pi_{k\Lambda})}(t)dt. \tag{3.129}$$

Using integration by parts, one has

$$\frac{d}{dt}P^{(\Pi_{k\Lambda})}(t) = \frac{1}{2}\sum_{X\in\mathcal{C}_{\Pi_{k\Lambda}}} \Delta_X^2\left(1 - \langle q_X\rangle_t^{(\Pi_{k\Lambda})}\right). \tag{3.130}$$

The translation symmetry over the torus implies the equality

$$2\sum_{X\subset\mathcal{C}_{\Pi_{k\Lambda}}} \Delta_X^2 = k^d \sum_{X\subset\mathcal{C}_{\Pi_\Lambda}} \Delta_X^2 \tag{3.131}$$

and by consequence

$$T_\Lambda^{(\phi)} = -\frac{1}{4}\lim_{k\to\infty}\sum_{X\subset\mathcal{C}_{\Pi_\Lambda}} \Delta_X^2\int_0^1\left(1 - \langle q_X\rangle_t^{(\Pi_{k\Lambda})}\right)dt. \tag{3.132}$$

The bound (3.119) immediately leads to (3.121). □

Theorem 3.32 (Integral representation for $T_\Lambda^{(\Pi)}$ and $T_\Lambda^{(\Pi^*)}$, Gaussian case) *For every* Λ *the symmetry of the Gaussian distribution implies*

$$T_\Lambda^{(\Pi)} = T_\Lambda^{(\Pi^*)}. \tag{3.133}$$

Consider the interpolating Hamiltonian (3.108) and let $\langle - \rangle_t^{(\Pi_\Lambda)}$ *be its quenched state. Then*

$$T_\Lambda^{(\Pi)} = T_\Lambda^{(\phi)} + \frac{1}{2}\sum_{X\subset\mathcal{C}_{\Pi_\Lambda}} \Delta_X^2\int_0^1\left(1 - \langle q_X\rangle_t^{(\Pi_\Lambda)}\right)dt. \tag{3.134}$$

In particular, if the condition (3.119) is fulfilled, then the quantity

$$\tau_\Lambda^{(\Pi)} = \frac{T_\Lambda^{(\Pi)}}{|\Lambda|^\alpha} \tag{3.135}$$

admits the bounds

$$\tau_\Lambda^{(\Phi)} \leqslant \tau_\Lambda^{(\Pi)} \leqslant \frac{\bar{c}}{2}. \tag{3.136}$$

Proof The proof is a simple adaptation of the one used for Theorem 3.27. In fact, by (3.113),

$$P_\Lambda^{(\Pi)} - P_\Lambda^{(\phi)} = T_\Lambda^{(\Pi)} - T_\Lambda^{(\phi)}. \tag{3.137}$$

The theorem comes from using the same interpolation scheme of Theorem 3.27, using the condition (3.119). □

Example 3.33 The same computation that was made in Example 3.28 shows that, in the d-dimensional Edwards–Anderson model (cf. Eq. (1.1)), condition (3.119) is fulfilled with $\bar{c} = 1$. In fact,

$$\frac{1}{2dL^{d-1}} \sum_{X \in \mathcal{C}_{\Pi_\Lambda}} \Delta_X^2 = 1. \tag{3.138}$$

Remark 3.34 The integral formulas for the surface pressure show that it has the correct size when the suitable condition (3.119) is fulfilled. We have seen that the Edwards–Anderson model does indeed fulfill the condition. The previous theorems are in fact a bound on the size of the surface pressure. It could, in principle, happen that the overlap expectations in (3.118) and (3.134) approach the value 1 in the thermodynamic limit. Although such a situation is not expected to hold either in the mean-field picture (Mézard *et al.* (1987)) or in the droplet one (Fisher and Huse (1986)), one can check that a rigorous analysis can be done at high temperatures where the behavior of the overlap expectations can be estimated in terms of the small parameter β (Contucci and Graffi (2004b)). This is the content of the following theorem.

Theorem 3.35 (High temperature surface pressure) *Considering the Edwards–Anderson model with Hamiltonian (1.1) in d dimensions at temperature β, then*

1. *there exist $\overline{\beta}$ and $C > 0$ depending only on d, such that for all $\beta \leqslant \overline{\beta}$*

$$\frac{\tau_\Lambda^{(\phi)}}{\beta^2} \leq -C < 0; \tag{3.139}$$

2. *for any $\varepsilon > 0$ there exists $\beta^{(\varepsilon)} > 0$, such that for all $\beta \leqslant \beta^{(\varepsilon)}$*

$$\frac{\tau_\Lambda^{(\phi)}}{\beta^2} \leqslant -\frac{1}{4}(1-\varepsilon), \tag{3.140}$$

and equivalently

$$\frac{\tau_\Lambda^{(\Pi)}}{\beta^2} \geqslant \frac{1}{4}(1-\varepsilon), \tag{3.141}$$

uniformly in Λ.

Proof The result is obtained by cluster expansion methods suitably adapted to the Gaussian random variables interactions (see, for instance, Berretti (1985)). The only non-zero contributions to the surface pressure integral representation come from the subsets X that are nearest neighbors (we will refer to them as b, to denote bonds). The expansion tells us that, regardless of the boundary conditions, each $\langle q_b \rangle_t^{(\Pi_{k\Lambda})}$ is small for small β, and definitely away from 1. Applying Proposition 1 of Berretti (1985) to our problem, we may write

$$\langle q_b \rangle_t^{(\Pi_{k\Lambda})} = A_{k\Lambda}(b, \beta^2, t)\beta^2 + C_{k\Lambda}(b, \beta^2, t), \tag{3.142}$$

where:

1. for every ε we may choose
$$|C_{k\Lambda}(b, \beta^2, t)| \leqslant \frac{\varepsilon}{2}, \tag{3.143}$$
uniformly in all the variables; and
2. $A_{k\Lambda}(b, \beta^2, t)$ is bounded uniformly in (Λ, t) and is analytic in β for $\beta < \beta_0$, where β_0 depends only on the dimension d and not on Λ or t. We remark that the parity of the Gaussian variables yields the parity in β of each thermodynamic function, so that the odd powers of the cluster expansion vanish.

After integrating in t, we take the $k \to \infty$ limit of the previous relation (which exists by Theorem 1 of Berretti (1985) if $\beta < \beta_0$), and sum over all bonds in the boundary $\partial\Lambda$. We obtain:

$$\tau_\Lambda^{(\phi)} = -\frac{\beta^2}{4}\left[1 - (A_\Lambda(\beta^2)\beta^2 + C_\Lambda(\beta^2))\right], \tag{3.144}$$

with

$$C_\Lambda(\beta^2) = \lim_{k\to\infty} \int_0^1 dt \frac{1}{|\partial\Lambda|} \sum_{b\in\partial\Lambda} C_{k\Lambda}(b, \beta^2, t), \tag{3.145}$$

and

$$A_\Lambda(\beta^2) = \lim_{k\to\infty} \int_0^1 dt \frac{1}{|\partial\Lambda|} \sum_{b\in\partial\Lambda} A_{k\Lambda}(b, \beta^2, t). \tag{3.146}$$

We remind the reader that the multiplicative β^2 factor in (3.144) comes from the fact that the potential has interaction coefficients βJ whose variance is β^2. From (3.143) we derive the bound $|C_\Lambda| \leqslant \varepsilon/2$. On the other hand, since the correlation is bounded by 1, $|\langle q_b \rangle_t^{(\Pi_{k\Lambda})}| \leqslant 1$, and $A_{k\Lambda}(b, \beta^2, t)$ is bounded uniformly in (Λ, t), $K > 0$ independent of Λ such that

$$|A_\Lambda(\beta^2)| < K. \tag{3.147}$$

Hence there is a $\overline{\beta} > 0$ such that the quantity $|\beta^2 A_\Lambda(\beta^2)| < C_1 < 1 - \varepsilon/2$ if $\beta < \overline{\beta}$, uniformly in Λ. Hence, by (3.144), we get the existence of $C > 0$ independent of

Λ such that

$$\frac{\tau_\Lambda^{(\phi)}}{\beta^2} < -C < 0. \tag{3.148}$$

This proves assertion (1). To prove assertion (2), we remark that, given $\varepsilon > 0$, by (3.147) we can always choose $\beta(\varepsilon)$ in such a way that

$$|A_\Lambda(\beta^2)\beta^2| \leqslant \varepsilon/2 \tag{3.149}$$

uniformly with respect to Λ if $\beta < \beta(\varepsilon)$. Hence by (3.144) we can conclude

$$\frac{\tau_\Lambda^{(\phi)}}{\beta^2} \leq -\frac{1}{4}(1-\varepsilon) \tag{3.150}$$

if $\beta < \beta(\varepsilon)$. The proof of (3.141) is completely analogous. □

Remark 3.36 The results just illustrated are for Gaussian interactions. For non-Gaussian interactions one can try to generalize the integration by parts formula (see Talagrand (2002) and section 4.2 of Guerra and Toninelli (2002)): if J_X is for all X an non-Gaussian even random variable with a finite fourth moment, the integration by parts (2.46) is replaced by

$$\mathbb{E}\left[J_X F(J)\right] = \mathbb{E}\left[J_X^2 F'(J)\right] - \frac{1}{4}\mathbb{E}\left[|J_X| \int_{-|J_X|}^{|J_X|} (J_X^2 - x^2) F'''(x)dx\right]. \tag{3.151}$$

When used within the previous results it generates a correction of order $O(\sqrt{|\Lambda|})$. When the pressure per particle is taken the correction vanishes in the thermodynamic limit. As for the surface pressure, the correction gives a finite contribution for $d = 2$ and a vanishing one for $d > 2$ so that the size of the first correction to the pressure is of surface size for all $d \geqslant 2$.

A different approach, which also turns out to be less demanding as far as the conditions of the coupling distributions are concerned, is to avoid integration by parts and use instead the interpolation with the linear power t instead of $\sqrt{t}$ as used in Theorem 3.2. This has been done in Contucci and Lebowitz (2007, 2010) which we follow here.

Theorem 3.37 (Integral representation for $T_\Lambda^{(\phi)}$, general centered interaction – classical and quantum) *For the model (3.72) with general centered interactions with $\mathbb{E}\left[J_X^2\right] = \Delta_X^2$, the surface pressure per unit surface admits the representation*

$$T_\Lambda^{(\phi)} = -\frac{1}{2}\lim_{k\to\infty} \sum_{X\subset \mathcal{C}_{\Pi_\Lambda}} \int_0^1 \langle J_X \sigma_X \rangle_t^{(\Pi_{k\Lambda})} dt. \tag{3.152}$$

In particular, if the following condition is fulfilled for some $0 < \alpha < 1$

$$\sup_{\Lambda} \frac{1}{|\Lambda|^{\alpha}} \sum_{X \in \mathcal{C}_{\Pi_\Lambda}} \Delta_X \leqslant \bar{c} < \infty \tag{3.153}$$

then the quantity

$$\tau_\Lambda^{(\phi)} = \frac{T_\Lambda^{(\phi)}}{|\Lambda|^{\alpha}} \tag{3.154}$$

admits the bounds

$$-\frac{\bar{c}}{2} \leqslant \tau_\Lambda^{(\phi)} \leqslant 0. \tag{3.155}$$

Proof We adapt the interpolation introduced in the proof of Theorem 3.31 by replacing $\sqrt{t_X}$ with t_X which immediately leads to (3.152). The bounds (3.155) come from the assumption (3.153) and the Schwartz inequality $\mathbb{E}\,[J_X \omega_t(\sigma_X)] \leqslant \Delta_X$ where we assume, without loss of generality, operators bounded in norm by 1. □

Remark 3.38 We notice that the condition (3.153) can be relaxed to the square summable one by using the bounds shown in Lemma 3.17. Lemma 3.18 implies that the same result can also be obtained in the case in which the interactions do not have a second moment but only a bounded and summable $\mathbb{E}\,[|J_X|]$.

3.9 Complete theory on the Nishimori line

As explained in Section 2.6, the Nishimori line is a special spin glass where the interaction parameters mean and variance are fixed to be identical. We could argue that the existence of the thermodynamic limit follows from the general Theorem 3.25 (where nevertheless the property of monotonicity is lost) and that the behavior of the surface pressure may be deduced by suitably adapting Theorem 3.37 to non-centered interactions. However, we take a different path because of the stronger properties that we have proved in the Nishimori line, which will allow us to deduce better limiting theorems – indeed, the most complete result for both pressure and surface pressure. The improvement with respect to the general case is mostly due to the availability of the correlation inequality of the second order (2.95) which implies monotonicity of the correlation functions with respect to the volume.

Theorem 3.39 (Thermodynamic limit on the Nishimori line) *On the Nishimori line defined by*

$$H_\Lambda(\sigma) = -\sum_{X \subset \Lambda} J_X \sigma_X, \tag{3.156}$$

with

$$\mathbb{E}[J_X] = \mathbb{E}\left[(J_X - \gamma_X)^2\right] = \gamma_X, \tag{3.157}$$

if the condition

$$\sup_\Lambda \frac{1}{|\Lambda|} \sum_{X \subset \Lambda} \gamma_X \leqslant \bar{c} < +\infty \tag{3.158}$$

is fulfilled, then the thermodynamic limit for pressure exists (in the sense of Fisher) as

$$\lim_{\Lambda \nearrow \mathbb{Z}^d} \frac{P_\Lambda}{|\Lambda|} = \sup_\Lambda \frac{P_\Lambda}{|\Lambda|} = p, \tag{3.159}$$

and is independent from free, periodic or antiperiodic boundary conditions. Moreover, the limit for the correlation functions exists (in the sense of Fisher)

$$\lim_{\Lambda \nearrow \mathbb{Z}^d} \langle \sigma_X \rangle_\Lambda = \sup_\Lambda \langle \sigma_X \rangle_\Lambda = \langle \sigma_X \rangle. \tag{3.160}$$

Proof The proof for the pressure proceeds as in Theorem 3.1, i.e. using superadditivity and boundedness. To establish the first property, the peculiar structure of the Nishimori line suggests an interpolation scheme on the Gaussian measure instead of the one on the Hamiltonian. In fact the pressure on the Nishimori line $P_\Lambda(\gamma)$ depends on the parameters $\gamma = \{\gamma_X\}_{X \subset \Lambda}$ only through the Gaussian distribution. The interpolation is defined as

$$\gamma_X(t) = \begin{cases} t\gamma_X, & \text{if } X \in \mathcal{C}_\Lambda, \\ \gamma_X, & \text{otherwise} \end{cases} \tag{3.161}$$

(see Lemma 3.2). By inspection, one sees that $P_\Lambda(\gamma(1)) = P_\Lambda(\gamma)$ and $P_\Lambda(\gamma(0)) = \sum_{s=1}^n P_{\Lambda_s}(\gamma)$, since they correspond to the singular Gaussian integration for mean and variance identically equal to zero on the corridors. An immediate computation which uses Eq. (2.94) gives

$$\frac{\partial P_\Lambda(\gamma(t))}{\partial t} = \sum_{X \in \mathcal{C}_\Lambda} \frac{\gamma_X}{2} (\mathbb{E}[\omega(\sigma_X)] + 1) \geq 0, \tag{3.162}$$

from which superadditivity follows. The boundedness comes from the elementary annealed bound on the pressure (see Lemma 3.4) and gives

$$P_\Lambda \leqslant |\Lambda| \left(\ln 2 + \frac{3\bar{c}}{2} \right) \tag{3.163}$$

from which the existence of the limit for the pressure follows in the Fisher sense. The proof for the correlation functions is based on the correlation inequality (2.95).

Given $\Lambda \subset \bar{\Lambda}$, we introduce the interpolation scheme

$$\gamma_X(t) = \begin{cases} \gamma_X, & \text{if } X \subset \Lambda, \\ t\gamma_X, & \text{otherwise.} \end{cases} \tag{3.164}$$

With such dependence on t we notice that

$$\langle \sigma_X \rangle_{\bar{\Lambda}}(\gamma(1)) = \langle \sigma_X \rangle_{\bar{\Lambda}}, \tag{3.165}$$

and

$$\langle \sigma_X \rangle_{\bar{\Lambda}}(\gamma(0)) = \langle \sigma_X \rangle_{\Lambda}. \tag{3.166}$$

By the fundamental theorem of calculus

$$\langle \sigma_X \rangle_{\bar{\Lambda}} - \langle \sigma_X \rangle_{\Lambda} = \int_0^1 dt \frac{d}{dt} \langle \sigma_X \rangle_{\bar{\Lambda}}(\gamma(t)), \tag{3.167}$$

where, by (2.95),

$$\frac{d}{dt} \langle \sigma_X \rangle_{\bar{\Lambda}}(\gamma(t)) = \sum_{Y \subseteq \bar{\Lambda}, Y \not\subseteq \Lambda} \frac{\gamma_Y}{2} \mathbb{E}\left[[\omega(\sigma_X \sigma_Y) - \omega(\sigma_X)\omega(\sigma_Y)]^2 \right] \geqslant 0. \tag{3.168}$$

Since the correlation functions are bounded by 1, the proved monotonicity implies the existence of their thermodynamic limit for a sequence of growing volumes in the Fisher sense. □

We now move to the study of the surface pressure. For general non-zero average interactions, one can obtain an integral representation for the surface pressure by suitably generalizing Theorem 3.37. What is not available is control of the sign of the surface pressure – since for general non-zero averages there is no type I correlation inequality. Although the Nishimori line is a non-zero average case, here we have a type I inequality to control the sign of the surface pressure, as well as a type II inequality which allows complete control in the thermodynamic limit of the adjacency surface pressure.

Definition 3.40 (Adjacency surface pressure) Consider the box $\Lambda = [0, 2L]^d$ composed of boxes $(\Lambda^{(s)})_{s=1,\dots,2^d}$ of volumes L^d. We define the *free boundary condition corridor* $\mathcal{C}_L$ to be the set

$$\mathcal{C}_L = \{X \subset \Lambda : X \not\subseteq \Lambda^{(s)} \quad \text{for } s = 1, \dots 2^d\}. \tag{3.169}$$

The difference between the quenched pressure of the box Λ (with free boundary conditions) and the pressures of the composing sub-boxes $\Lambda^{(s)}$ is called the *adjacency surface pressure* T_L:

$$T_L = P_{2L} - 2^d P_L. \tag{3.170}$$

Theorem 3.41 (Adjacency surface pressure existence) *On the Nishimori line, the adjacency pressure T_L is always positive. For a sequence of γ_X on growing boxes of side L with free boundary conditions satisfying the surface bound*

$$\sup_L \frac{1}{L^{d-1}} \sum_{X \in \mathcal{C}_L} \gamma_X \leqslant \bar{c} < \infty, \tag{3.171}$$

the adjacency pressure grows proportionally to the surface size L^{d-1}. If it exists

$$\lim_{L\to\infty} \frac{1}{L^{d-1}} \sum_{X \in \mathcal{C}_L} \gamma_X = c_1, \tag{3.172}$$

then the ratio T_L/L^{d-1} exists in the thermodynamic limit and admits the representation

$$\lim_{L\to\infty} \frac{T_L}{L^{d-1}} = \frac{1}{2}\left(\int_0^1 \langle m_{\mathcal{C}} \rangle_t dt + c_1\right), \tag{3.173}$$

where

$$\langle m_{\mathcal{C}} \rangle_t = \lim_{L\to\infty} \frac{1}{L^{d-1}} \sum_{X \in \mathcal{C}_L} \gamma_X \langle \sigma_X \rangle_t. \tag{3.174}$$

Example 3.42 For the Edwards–Anderson model (Definition 1.1) one has

$$\lim_{L\to\infty} \frac{T_L}{L^{d-1}} = \gamma d 2^{d-2}\left(1 + \int_0^1 \langle \sigma_1 \sigma_2 \rangle_t \, dt\right) \tag{3.175}$$

where the two spins σ_1 and σ_2 are two *fixed* nearest neighbors of two adjacent boxes far away from the outer boundaries.

Proof Consider the interpolating pressure $P(\gamma(t))$ defined by

$$\gamma_X(t) = \begin{cases} t\gamma_X, & \text{if } X \in \mathcal{C}_L, \\ \gamma_X, & \text{otherwise.} \end{cases} \tag{3.176}$$

One sees that

$$T_L = P(\gamma(1)) - P(\gamma(0)) = \int_0^1 dt \frac{d}{dt} P(\gamma(t)). \tag{3.177}$$

Since

$$\frac{d}{dt}P(\gamma(t)) = \sum_{X \in C_L} \frac{\gamma_X}{2}(\mathbb{E}\,[\omega(\sigma_X)]_t + 1) \geq 0, \tag{3.178}$$

it follows that $T_L \geqslant 0$ and because of the surface bound (3.171) one has the growth of T_L proportional to L^{d-1}. The representation theorem comes from observing that

$$T_L = \int_0^1 dt \sum_{X \in C_L} \frac{\gamma_X}{2}(\mathbb{E}\,[\omega(\sigma_X)]_t + 1). \tag{3.179}$$

The term c_1 in Eq. (3.173) follows from hypothesis (3.172) and the existence of the term (3.174) is a consequence of the monotonicity of each $\langle\sigma_X\rangle_t$ (due to the type II inequality) and of the surface bound (3.171). □

In general, for non-zero mean interaction, one can obtain a representation theorem for the surface pressure and control on the size – provided that the necessary technical conditions on the mean interaction are met. However there is no control on the sign of the surface pressure. Nevertheless, on the Nishimori line, the correlation inequality of type I (2.94) allows recovery of control of the sign through the proper interpolation introduced in Theorem 3.39. Therefore we have the following theorems (we omit the proofs because they can be derived in a straightforward way in analogy with Theorems 3.31 and 3.32).

Theorem 3.43 (Integral representation for $T_\Lambda^{(\phi)}$ on the Nishimori line) *With reference to the definition given in (3.115) and using the Nishimori line interpolation on the* γ *s, the surface pressure with free boundary conditions admits the representation*

$$T_\Lambda^{(\phi)} = -\frac{L^{d-1}}{2}\int_0^1 (\langle m_{C_{\Pi_\Lambda}}\rangle_t + c_1(\Lambda))dt, \tag{3.180}$$

where

$$\langle m_{C_{\Pi_\Lambda}}\rangle_t = \lim_{k\to\infty} \sum_{X \subset C_{\Pi_\Lambda}} \gamma_X \mathbb{E}\left[\omega_t^{(\Pi_{k\Lambda})}(\sigma_X)\right] \tag{3.181}$$

and

$$c_1(\Lambda) = \sum_{X \subset C_{\Pi_\Lambda}} \gamma_X. \tag{3.182}$$

In particular, it follows by inspection and inequality (2.94) that the surface pressure with free boundary conditions is always negative and, if the following bound is satisfied

$$\sup_L \frac{1}{L^{d-1}} \sum_{X \in \mathcal{C}_{\Pi_\Lambda}} \gamma_X \leqslant \bar{c} < \infty, \tag{3.183}$$

then it is of surface size with bounds

$$-L^{d-1}\bar{c} \leqslant T_\Lambda^{(\phi)} \leqslant 0. \tag{3.184}$$

Theorem 3.44 (Integral representation for $T_\Lambda^{(\Pi)}$ on the Nishimori line) *Consider the Nishimori line interpolation on the γ's and let $\langle - \rangle_t^{(\Pi_\Lambda)}$ be its quenched state. Then*

$$T_\Lambda^{(\Pi)} = T_\Lambda^{(\phi)} + \frac{1}{2} \sum_{X \subset \mathcal{C}_{\Pi_\Lambda}} \gamma_X \int_0^1 \left(1 + \langle \sigma_X \rangle_t^{(\Pi_\Lambda)}\right) dt. \tag{3.185}$$

In particular, if the condition (3.183) is fulfilled, then the quantity

$$\tau_\Lambda^{(\Pi)} = \frac{T_\Lambda^{(\Pi)}}{L^{d-1}} \tag{3.186}$$

admits the bounds

$$\tau_\Lambda^{(\phi)} \leqslant \tau_\Lambda^{(\Pi)} \leqslant c. \tag{3.187}$$

Remark 3.45 It is interesting to observe that on the Nishimori line of the Edwards–Anderson model, the surface pressure fulfills the inequalities $T_\Lambda^{(\phi)} \leqslant 0$ – the same as in the ferromagnetic case. This fact is totally nontrivial a priori because the interactions have no definite sign in our model and the Griffiths inequalities do not hold in the standard form. Nevertheless, the positivity of the average of the interactions is enough to guarantee that the sign of the surface pressure persists in the disordered case.

3.10 Thermodynamical limit for mean-field models, REM and GREM

When specialized to volume-dependent interaction distribution, general spin glass models (introduced in Definition 1.4) play a very important role in spin glass theory – including the mean-field case like the Sherrington–Kirkpatrick model (Definition 1.3), the Random Energy Model (introduced in Derrida (1980)), and its Generalized version (introduced in Derrida (1985)).

Because of the volume dependence in the distribution of the interaction variables J_X, the interpolation strategy introduced for finite-dimensional models cannot be extended trivially to mean-field models. This is the motivation to deal with mean-field models in a different way, stressing the fact that the whole interpolation method came from the brilliant work by Guerra and Toninelli (2002), who solved the longstanding conjecture of existence of the thermodynamic limit for the Sherrington–Kirkpatrick system.

We start the mean-field approach with the following.

Theorem 3.46 (Comparison of Gaussian processes) *Let* $\{\xi_i\}_{i=1,\dots,n}$ *and* $\{\eta_i\}_{i=1,\dots,n}$ *be two centered independent Gaussian processes with covariances* $C^{(\xi)}(i,j) = \mathbb{E}\left[\xi_i\xi_j\right]$ *and* $C^{(\eta)}(i,j) = \mathbb{E}\left[\eta_i\eta_j\right]$*, respectively. If the following inequalities are fulfilled*

$$C^{(\xi)}(i,j) \leqslant C^{(\eta)}(i,j) \qquad \forall (i,j) \tag{3.188}$$

$$C^{(\xi)}(i,i) = C^{(\eta)}(i,i) \qquad \forall i \tag{3.189}$$

then

$$\mathbb{E}\left[\ln\sum_{i=1}^{n}\exp\xi_i\right] \geqslant \mathbb{E}\left[\ln\sum_{i=1}^{n}\exp\eta_i\right]. \tag{3.190}$$

Proof For $t \in [0,1]$, consider the interpolating function

$$\mathbb{E}\left[\ln\sum_{i=1}^{n}\exp\sqrt{t}\xi_i + \sqrt{1-t}\eta_i\right], \tag{3.191}$$

its value for $t = 0$

$$\mathbb{E}\left[\ln\sum_{i=1}^{n}\exp\eta_i\right], \tag{3.192}$$

its value for $t = 1$

$$\mathbb{E}\left[\ln\sum_{i=1}^{n}\exp\xi_i\right], \tag{3.193}$$

and its derivative

$$\mathbb{E}\left[\sum_{i=1}^{n}\left(\frac{1}{2\sqrt{t}}\xi_i - \frac{1}{2\sqrt{1-t}}\eta_i\right)\frac{\exp\sqrt{t}\xi_i + \sqrt{1-t}\eta_i}{\sum_{i=1}^{n}\exp\sqrt{t}\xi_i + \sqrt{1-t}\eta_i}\right]. \tag{3.194}$$

The sign of the derivative can be found by integrating by parts. Indeed, using the generalized Wick's theorem (1.19) we find that the derivative is equal to

$$\frac{1}{2}\mathbb{E}\left[\sum_{i=1}^{n}((C^{(\xi)}(i,i)-C^{(\xi)}(i,j))-(C^{(\eta)}(i,i)-C^{(\eta)}(i,j)))\right.$$

$$\left.\times\frac{\exp\sqrt{t}\xi_i+\sqrt{1-t}\eta_i}{\sum_{i=1}^{n}\exp\sqrt{t}\xi_i+\sqrt{1-t}\eta_i}\frac{\exp\sqrt{t}\xi_j+\sqrt{1-t}\eta_j}{\sum_{i=1}^{n}\exp\sqrt{t}\xi_i+\sqrt{1-t}\eta_i}\right], \tag{3.195}$$

which is clearly positive because of the hypotheses (3.188) and (3.189). The claimed inequality follows by applying the fundamental theorem of calculus. □

Remark 3.47 The same strategy as used in the previous proof allows us to compare expectations of a larger class of functions of two Gaussian processes which are more general than the "pressures". Indeed, a result by Kahane states that for a measurable function f it suffices to have a positive second cross-derivative to obtain $\mathbb{E}[f(\eta)] \geqslant \mathbb{E}[f(\xi)]$ under the same hypothesis of the theorem (Kahane (1986)). A particular case was considered by Slepian (see Bovier (2006); Slepian (1962)) for the function $f(x) = \max\{x_1, \ldots, x_n\}$.

To prove the existence of the limit in mean-field models, we follow the formulation given in Contucci *et al.* (2003).

Theorem 3.48 (Pressure superadditivity, mean-field) *Consider a mean-field model with Hamiltonian $H_N(\sigma)$ with covariance $C_N(\sigma,\tau) = Nc_N(\sigma,\tau)$. Without loss of generality we assume $c_N(\sigma,\sigma) = 1$. For each $\sigma \in \Sigma_N = \{+1,-1\}^N$ let π_1 and π_2 be the two canonical projections over the two subsets Σ_{N_1} and Σ_{N_2}, generated by a partition $\mathcal{P}$ of the coordinates $(\sigma_1, \ldots, \sigma_N)$ into a subset of N_1 coordinates, and into a complementary set of N_2 coordinates: $N_1 + N_2 = N$, $\Sigma_N = \Sigma_{N_1} \times \Sigma_{N_2}$, $\pi_1 \otimes \pi_2 = 1_{\Sigma_N}$.*

Let the covariance matrices C_N fulfill the condition:

$$c_N(\sigma,\tau) - \frac{N_1}{N}c_{N_1}(\pi_1(\sigma),\pi_1(\tau)) - \frac{N_2}{N}c_{N_2}(\pi_2(\sigma),\pi_2(\tau)) \leqslant 0 \tag{3.196}$$

for every $N \geqslant \tilde{N}$, every $(\sigma,\tau) \in \Sigma_N \times \Sigma_N$, and every decomposition $N_1 + N_2 = N$. Then the thermodynamical limit of the quenched pressure exists and is given by

$$\lim_{N\to\infty}\frac{1}{N}\mathbb{E}\left[\log \mathcal{Z}_N(\beta)\right] = \sup_N \frac{1}{N}\mathbb{E}\left[\log \mathcal{Z}_N(\beta)\right]. \tag{3.197}$$

Proof The proof can be deduced from the previous theorem with the identification $\xi = H_N$ and $\eta = \tilde{H}_{N_1} + \bar{H}_{N_2}$, where $\tilde{H}_{N_1}$ and $\bar{H}_{N_2}$ are independent copies of the Hamiltonian for system sizes N_1 and N_2 respectively.

The boundedness follows from the elementary Jensen inequality

$$\mathbb{E}\left[\log \mathcal{Z}\right] \leqslant \log(\mathbb{E}\left[\mathcal{Z}\right]) = N\left(\log 2 + \frac{\beta^2}{2}\right). \tag{3.198}$$

□

Remark 3.49 The result (3.197) can be extended to the almost everywhere convergence of free energy density, internal energy, and ground state energy with elementary probability methods (see Guerra and Toninelli (2002)). In particular, the pressure per particle concentrates around its mean value as described in Theorem 3.10.

The following are a series of examples in the context of mean-field models that can be treated with the above theorem.

Example 3.50 The *Sherrington–Kirkpatrick model* (Definition 1.3) given by the covariance

$$c_N(\sigma, \tau) = [q_N(\sigma, \tau)]^2 \tag{3.199}$$

where, as usual

$$q_N(\sigma, \tau) := \frac{1}{N}\sum_{k=1}^{N} \sigma_k \tau_k \tag{3.200}$$

is the overlap between the σ and τ spin configurations, fulfills the condition (3.196) as shown in Guerra and Toninelli (2002), using

$$q_N(\sigma, \tau) - \frac{N_1}{N} q_{N_1}(\pi_1(\sigma), \pi_1(\tau)) - \frac{N_2}{N} q_{N_2}(\pi_2(\sigma), \pi_2(\tau)) = 0, \tag{3.201}$$

and the convexity property of the function $x \mapsto x^2$.

Example 3.51 For *p-spin* models defined by the covariance

$$c_N(\sigma, \tau) = [q_N(\sigma, \tau)]^p, \tag{3.202}$$

condition (3.196) is fulfilled by convexity for *even p*.

Example 3.52 The *Derrida random energy model*. Here the model is specified by the covariance

$$c_N(\sigma, \tau) = \delta(\sigma, \tau) \tag{3.203}$$

where $\delta(\sigma, \tau)$ denotes the Kronecker delta. Condition (3.196) is verified because it becomes

$$\delta_{\sigma,\sigma'} \leq \frac{N_1}{N}\delta_{\pi_1(\sigma),\pi_1(\sigma')} + \frac{N_2}{N}\delta_{\pi_2(\sigma),\pi_2(s')}. \tag{3.204}$$

In fact if $\sigma = \sigma'$ the previous formula is an identity. If $\sigma \neq \sigma'$ the lefthand side is zero but the righthand side is not always zero. Let us take for instance

$$\sigma = (+,+)$$
$$\sigma' = (+,-)$$
$$\pi_1(+,+) = +$$
$$\pi_1(+,-) = +$$
$$\pi_2(+,+) = +$$
$$\pi_2(+,-) = -.$$

In this case the lefthand side is zero and the righthand side is $1/2$.

Example 3.53 The *Derrida–Gardner generalized random energy model.* We consider rooted trees with a fixed number n of layers and 2^N leaves. One tree is identified by the sequence $\{K_i\}_{i=1}^n$, i.e. 2^{K_i} furcations occur at each level i. The space of trees is *partially ordered* by the notion of *inclusion*: we say that $\mathcal{T}' \subset \mathcal{T}''$ when $K_i^{'} \leqslant K_i^{''}$. Considering $\mathcal{T} \supset \mathcal{T}_1$, the relation $K_i = K_i(1) + K_i(2)$ uniquely defines the complementary tree $\mathcal{T}_2$ such that

$$\mathcal{T} = \mathcal{T}_1 \otimes \mathcal{T}_2. \tag{3.205}$$

Let us show that the condition (3.196) is fulfilled by all triples verifying (3.205). For a given tree $\mathcal{T}$, the GREM is defined by its covariance

$$\begin{aligned} c_N(\sigma,\tau) = {} & a_1\delta(\sigma_1,\tau_1)\cdots\delta(\sigma_{K_1},\tau_{K_1}) \\ & + a_2\delta(\sigma_1,\tau_1)\cdots\delta(\sigma_{K_1+K_2},\tau_{K_1+K_2}) \\ & + \cdots \\ & + a_n\delta(\sigma_1,\tau_1)\cdots\delta(\sigma_{K_1+\cdots+K_n},\tau_{K_1+\cdots+K_n}), \end{aligned} \tag{3.206}$$

with $a_i > 0$ and $\sum_{i=1}^n a_i = 1$. In other words, considering two configurations $\sigma = (\sigma_1, \sigma_2, \ldots, \sigma_N)$ and $\tau = (\tau_1, \tau_2, \ldots, \tau_N)$, one has

$$c_N(\sigma,\tau) = \sum_{i=1}^m a_i \tag{3.207}$$

where m is such that $\sigma_i = \tau_i \forall i \in [1, K_1 + \cdots + K_m]$ and $\exists i \in [K_1 + \cdots + K_m + 1, K_1 + \cdots + K_{m+1}]$ s.t. $\sigma_i \neq \tau_i$.

Considering the decomposition relative to (3.205), $N = N_1 + N_2$ and $K_i = K_i(1) + K_i(2) \forall i = 1, \ldots, n$, we can use (at each level i) the basic REM inequality (3.204) (written here for a generic $j = j_1 + j_2$):

$$\delta(\sigma_1, \tau_1) \cdots \delta(\sigma_j, \tau_j) \leqslant \frac{N_1}{N} \delta(\sigma_1, \tau_1) \cdots \delta(\sigma_{j_1}, \tau_{j_1}) + \frac{N_2}{N} \delta(\sigma_{j_1+1}, \tau_{j_1+1}) \cdots \delta(\sigma_{j_1+j_2}, \tau_{j_1+j_2}). \quad (3.208)$$

Using the previous relation repeatedly, one easily verifies the condition (3.196).

Remark 3.54 In Jana and Rao (2006), an explicit example of the violation of (3.196) is illustrated if the condition (3.205) is not met.

4

Exact results for mean-field models

Abstract

Some of the recent mathematical results which established the Parisi solution of the Sherrington–Kirkpatrick model on firm ground are reviewed. We first introduce the stochastic processes that arise in the study of mean-field spin glasses, i.e. the Poisson point process with exponential intensity and the Poisson–Dirichlet distribution. We then show how these processes (and their extensions given by the Ruelle random probability cascades), combined with the interpolation technique, allow the (generalized) random energy model to be solved. In the last section, the replica symmetry breaking bound for the pressure of the Sherrington–Kirkpatrick model obtained by Guerra is presented. The bound, which shows that Parisi pressure is an upper bound for the Sherrington–Kirkpatrick model, is proved as a particular case of the more general upper bound given by the Aizenman–Sims–Starr extended variational principle. The Parisi expression is obtained when the variational bound is evaluated on the random probability cascades and on a hierarchical overlap structure. Some remarks on Talagrand's solution and open problems of the Sherrington–Kirkpatrick model close the chapter.

4.1 Introduction to the Poisson point process

We first introduce a *point process* on $E \subseteq \mathbb{R}$. This is obtained by distributing random points on the subset E of the real line, and introducing the counting measure which counts the random number of points which fall in a measurable interval. Denoting the random points by $\{X_i, i \geqslant 1\}$ and an open interval by $A = (a, b) \subseteq E$, we shall define the point process by the integer-valued variable

$$N(A) = \sum_{i \geqslant 1} \mathbb{1}_A(X_i) \tag{4.1}$$

where

$$\mathbb{1}_A(x) = \begin{cases} 1 & \text{if } x \in A \\ 0 & \text{if } x \notin A. \end{cases} \tag{4.2}$$

Thus $N(A)$ counts the *random* number of points in A. We define the *intensity* (or *mean measure*), μ, of the point process such that

$$\mu(A) = \mathbb{E}(N(A)), \tag{4.3}$$

where $\mathbb{E}$ denotes expectation over the randomness of the points. Thus $\mu(A)$ is the expected number of points in set A. We are now ready to define the *Poisson point process* on $E \subseteq \mathbb{R}$.

Definition 4.1 (Poisson point process) A point process on $E \subseteq \mathbb{R}$ with intensity μ is a Poisson point process (also called a Poisson random measure PRM(μ)) if the following two conditions are met:

1. $N(A)$ has Poisson distribution with parameter $\mu(A)$;
2. if $A_1, \ldots, A_n$ are disjoint intervals, then $N(A_1), \ldots, N(A_n)$ are independent random variables.

The standard way to construct a Poisson point process on a subset $E \subseteq \mathbb{R}$ with a given intensity μ is as follows. We will start with the case $\mu(E) < \infty$ and define the probability measure

$$\tilde{\mu}(dx) = \frac{\mu(dx)}{\mu(E)}. \tag{4.4}$$

An occurrence of the points of the Poisson point process with intensity measure μ is generated in two steps: first, the number of points is chosen by sampling a Poisson random variable with parameter $\mu(E)$; second, assuming n points are to be generated, the points $x_1, \ldots, x_n$ are taken to be a realization of the sequence of random variables $\{X_i\}_{i=1}^n$, which are i.i.d. with common distribution $\tilde{\mu}$. Thus, conditional on there being n points in the interval E, these points are distributed as n i.i.d. $\tilde{\mu}$-distributed random variables.

One can check that both properties in Definition 4.1 are satisfied. Let us check the first property. It is enough to compute the generating function

$$\begin{aligned}
\mathbb{E}\left(e^{-tN(A)}\right) &= \mathbb{E}\left(e^{-t\sum_i \mathbb{1}_A(X_i)}\right) \\
&= \sum_{n=0}^{\infty} \mathbb{E}\left(e^{-t\sum_{i=1}^{N(E)} \mathbb{1}_A(X_i)} | N(E) = n\right) \mathbb{P}(N(E) = n) \\
&= \sum_{n=0}^{\infty} \left[\mathbb{E}\left(e^{-t\mathbb{1}_A(X_1)}\right)\right]^n \mathbb{P}(N(E) = n) \\
&= \sum_{n=0}^{\infty} \left(\frac{e^{-t}\mu(A) + \mu(E) - \mu(A)}{\mu(E)}\right)^n \frac{\mu(E)^n}{n!} e^{-\mu(E)} \\
&= e^{\mu(A)(e^{-t}-1)},
\end{aligned}$$

which is recognized to be the generating function of a Poisson random variable with parameter $\mu(A)$. The second property is checked in an analogous way using generating functions and independence.

When $\mu(E) = +\infty$, we decompose E into disjoint sets $E_1, E_2, \ldots$ such that $E = \cup_k E_k$ and $\mu(E_k) < \infty$ for each k. Define $\mu_k(dx) = \mu(dx)\mathbb{1}_{E_k}(x)$ and let N_k be the independent Poisson point process on E_k, constructed as above. One can check that $N = \sum_k N_k$ is a Poisson point process on E since $\mu = \sum_k \mu_k$.

One important property of Poisson point processes is that when we transform the points X_i of the process with some mapping f, then the transformed points $Y_i = f(X_i)$ are the points of a new Poisson point process.

Proposition 4.2 (Transforming Poisson point processes) *Suppose that N is a Poisson point process with points $\{X_i\}$ on E and with intensity μ. Suppose $f : E \mapsto E'$ is a mapping with the property that the pre-image of any bounded set in E' is also bounded in E. Then the transformed point process $N' = N \circ f^{-1}$ is a Poisson point process with points $\{f(X_i)\}$ on E' and with intensity $\mu' = \mu \circ f^{-1}$.*

Proof For $A' \subseteq E'$, we need to verify the two properties:

1. $N'(A') \sim \text{Poisson}(\mu'(A'))$;
2. if $A'_1, \ldots, A'_n$ are disjoint intervals, then $N'(A'_1), \ldots, N(A'_n)$ are independent random variables.

We have

$$\begin{aligned}\mathbb{E}(e^{-tN'(A')}) &= \mathbb{E}\big(e^{-t\,N\circ f^{-1}(A')}\big)\\ &= \mathbb{E}\big(e^{-t\,N(A)}\big)\\ &= e^{\mu(A)(e^{-t}-1)}\\ &= e^{\mu\circ f^{-1}(A')(e^{-t}-1)}\\ &= e^{\mu'(A')(e^{-t}-1)}.\end{aligned}$$

It is easy to check the independence property since, if $A'_1, \ldots, A'_n$ are disjoint intervals, then so are $f^{-1}(A'_1), \ldots, f^{-1}(A'_n)$, from which $N'(A'_1), \ldots, N(A'_n)$ are independent random variables. □

4.2 Poisson point process with exponential intensity

There is a choice for the intensity measure μ that is particularly relevant for mean-field spin glasses. We will now explain this choice and discuss its relevant properties.

We consider the Poisson point process on the real line ($E = \mathbb{R}$) with intensity measure $\mu(dx) = e^{-x}dx$. Since the total mass is infinite ($\int_{-\infty}^{+\infty} e^{-x}dx = +\infty$), to construct the process we should apply the construction described in the previous section and write, for example, $\mathbb{R} = \cup_{i=-\infty}^{+\infty}[i, i+1)$.

Alternatively, we can proceed in the construction by considering a finite cutoff $c \in \mathbb{R}$, so that the total mass is finite, and then take the limit $c \to -\infty$ at the end of the construction. To follow such a procedure, we define the truncated measure μ_c as

$$\mu_c(dx) = \begin{cases} e^{-x}dx & \text{if } x \geqslant c \\ 0 & \text{if } x < c. \end{cases} \tag{4.5}$$

The total mass of the measure μ_c is finite

$$\mu_c(\mathbb{R}) = \int_c^{+\infty} e^{-x}dx = e^{-c}. \tag{4.6}$$

Thus, we can construct the points of the Poisson point process on $\mathbb{R}$ with intensity μ_c by throwing points at random on $\mathbb{R}$ according to the distribution

$$\tilde{\mu}_c(dx) = \begin{cases} e^{-(x-c)}dx & \text{if } x \geqslant c \\ 0 & \text{if } x < c, \end{cases} \tag{4.7}$$

where the number of points thrown is Poisson with parameter $\mu_c(\mathbb{R}) = e^{-c}$. The points thrown will appear only on $(c, +\infty)$. The points of the Poisson point process on $\mathbb{R}$ with intensity $e^{-x}dx$ will be the limiting set of points that is obtained when $c \to -\infty$. From now on we will always work with this construction. In the limit $c \to -\infty$, with probability 1, the process will have an infinite number of points.

The exponential Poisson point process will appear in the description of the quenched state of mean-field spin glasses. It fulfills a special invariance property that will be crucial in the study of the random energy model and of its generalized version, as well as in the Parisi expression of the Sherrington–Kirkpatrick (SK) free energy. We will now discuss this invariance property.

Consider a sequence of i.i.d. random variables. If we add those random variables to the points of the exponential Poisson point process, we obtain a new exponential Poisson point process. This new process can be constructed equivalently by starting from the old Poisson point process and adding a proper constant to each point. In other words, shifting all the points of a Poisson point process with exponential intensity by an independent random amount, is like shifting all its points by a deterministic amount. This is made precise by the following result.

Proposition 4.3 (Invariance property of the exponential Poisson point process) *For any* $0 < m < 1$, *suppose* $\{U_i\}_{i\geqslant 1}$ *is a sequence of i.i.d. random variables with* $\mathbb{E}(e^{mU_1}) < \infty$. *Let* $\{X_i\}_{i\geqslant 0}$ *be the points of a Poisson point process with intensity* $e^{-x}dx$. *Then the following equality in distribution holds:*

$$\left\{\frac{X_i}{m} + U_i\right\}_{i\geqslant 1} \overset{\mathcal{D}}{=} \left\{\frac{X_i}{m} + \frac{1}{m}\ln(\mathbb{E}(e^{mU_1}))\right\}_{i\geqslant 1}. \tag{4.8}$$

Proof We will work using the construction of the Poisson point process with a finite cutoff c (and then send the cutoff $c \to -\infty$ at the end) and prove the equivalent statement

$$\{X_i + mU_i\}_{i\geqslant 1} \overset{\mathcal{D}}{=} \{X_i + \ln(\mathbb{E}(e^{mU_1}))\}_{i\geqslant 1}. \tag{4.9}$$

Let $g(x)$ be the common density of the random variables $\{U_i\}_{i\geqslant 1}$. It follows then that $l(x) = \frac{1}{m}g(\frac{x}{m})$ is the density of the random variables $\{mU_i\}$. On the condition that there are k points, the truncated version of the exponential Poisson point process is a collection of k i.i.d. random variables $\{X_i\}_{i=1}^k$ with distribution (4.7), i.e. with density function

$$f_c(x) = \begin{cases} e^{-(x-c)} & \text{if } x \geqslant c \\ 0 & \text{if } x < c. \end{cases} \tag{4.10}$$

Therefore, the random variables $\{X_i + mU_i\}_{i=1}^k$ will be a collection of k i.i.d. random variables with density function $h(x)$ given by the convolution

$$
\begin{aligned}
h(x) &= \int dy\, f_c(y) l(x-y) \\
&= \int_c^{+\infty} dy\, e^{-(y-c)} \frac{1}{m} g\left(\frac{x-y}{m}\right) \\
&= e^{-(x-c)} \int_{-\infty}^{\frac{x-c}{m}} du\, e^{mu} g(u).
\end{aligned} \tag{4.11}
$$

On the other hand, an immediate computation shows that when we add a constant a to k points of the exponential Poisson point process, we obtain random variables $\{X_i + a\}_{i=1}^k$ with density

$$
\tilde{h}(x) = \begin{cases} e^a e^{-(x-c)} & \text{if } x \geqslant c+a \\ 0 & \text{if } x < c+a. \end{cases} \tag{4.12}
$$

Thus, if we choose

$$
a = \ln(\mathbb{E}(e^{mU_1})) \tag{4.13}
$$

we see that, in the limit $c \to -\infty$, the two sets of points will have the same density. □

The invariance property described above is at the root of the quasi-stationarity property of the competing particle system considered in Aizenman *et al.* (2007); (see also Ruzmaikina and Aizenman (2005); Arguin (2007)). In this context, the invariance of the *ranked* points of the exponential Poisson point process under a random reshuffling, given by the addition of a generic noise, is investigated.

Up to now we have considered *unordered* sequences of points. *Ordered* sequences are also interesting (indeed, in the next section we will discuss the random ranked discrete distributions that are obtained by ordering the points of a Poisson point process and normalizing). Here we will discuss the order statistics of the points of the Poisson point process with intensity $e^{-x}dx$. As usual, we will study the order statistics of the truncated process and then send the truncation to negative infinity.

We recall the definition of order statistics for a set of random variables. Suppose $X_1, X_2, \dots X_n$ are i.i.d. random variables having a common continuous distribution function. We define new random variables $X_{(1)}, X_{(2)}, \dots, X_{(n)}$, called the order statistics, by

$$
X_{(i)} = \text{the } i\text{th largest of the random variables } X_1, X_2, \dots X_n
$$

so that

$$X_{(1)} > X_{(2)} > \cdots > X_{(n)}.$$

One can prove the following result.

Proposition 4.4 (Order statistics of the exponential Poisson point process) *The largest point $X_{(1)}$ of the points of the Poisson point process with intensity $e^{-x}dx$ has the probability density of the Gumbel distribution, i.e.*

$$f_{X_{(1)}}(x) = e^{-x-e^{-x}}. \tag{4.14}$$

The k largest points $X_{(1)}, \ldots, X_{(k)}$ of the points of the Poisson point process with intensity $e^{-x}dx$ have a joint probability density given by

$$f_{X_{(1)},\ldots,X_{(k)}}(x_1,\ldots,x_k) = \begin{cases} e^{-x_1-x_2-\cdots-x_k-e^{-x_k}} & \text{if } x_1 > x_2 > \cdots > x_k \\ 0 & \text{otherwise.} \end{cases}$$

Proof We check the formula for the distribution of the maximum of the points and leave to the interested reader the verification of the joint distribution for the k largest points. We start from the truncated process with intensity μ_c and define $X_{(1)}$ to be the largest of its points. Denoting by M the total number of points, for a given $x > c$, we have

$$\begin{aligned}
\mathbb{P}(X_{(1)} \leqslant x) &= \mathbb{P}(\max(X_1, X_2, \ldots) \leqslant x) \\
&= \sum_{n=0}^{\infty} \mathbb{P}(\max(X_1,\ldots,X_M) \leqslant x | M = n)\, \mathbb{P}(M = n) \\
&= \sum_{n=0}^{\infty} \mathbb{P}(X_1 \leqslant x, \ldots, X_n \leqslant x)\, \mathbb{P}(M = n) \\
&= \sum_{n=0}^{\infty} \mathbb{P}(X_1 \leqslant x)^n\, \mathbb{P}(M = n) \\
&= \sum_{n=0}^{\infty} \left(\int_c^x e^{-(y-c)} dy \right)^n \frac{(e^{-c})^n}{n!} e^{-e^{-c}} \\
&= e^{\int_c^x e^{-y} dy - e^{-c}} \\
&= e^{-e^{-x}}.
\end{aligned}$$

Therefore

$$f_{X_{(1)}}(x) = \frac{d\mathbb{P}(X_{(1)} \leqslant x)}{dx} = e^{-x-e^{-x}}. \tag{4.15}$$

□

4.3 The Poisson–Dirichlet distribution

In mean-field spin glasses, a prominent role is played by some *random* probability measures on the integers which are known as Poisson–Dirichlet distributions. These random probability measures arise in a wide variety of contexts such as coalescence theory, population genetics, random permutations, random graphs, and number theory. In this section we will discuss their general properties and, in particular, their representation in terms of the exponential Poisson point process.

The Poisson–Dirichlet distribution, denoted by $\mathrm{PD}(m, \theta)$[1], is a probability distribution over discrete distributions on the integers (with ranked weights), depending on the two parameters $\{0 \leqslant m < 1,\ \theta > -m\}$. In other words, it is a probability distribution on the ordered infinite-dimensional simplex

$$\Delta_{(\infty)} = \left\{ \xi_1 \geqslant \xi_2 \geqslant \cdots \geqslant 0 : \ \sum_{i=1}^{\infty} \xi_i = 1 \right\}. \tag{4.16}$$

It was introduced by Kingman who considered the case $\{m = 0,\ \theta > 0\}$ and introduced the process $\mathrm{PD}(0, \theta)$ in relation to partition structures and random permutations (Kingman (1975)). Pitman and Yor (1997) defined the generalized two-parameter version $\mathrm{PD}(m, \theta)$. Ruelle independently defined the random probability cascade, a hierarchical model, whose 1-level marginals are given by $\mathrm{PD}(m, 0)$ with $0 < m < 1$, depending on the depth of the level in the tree. Ruelle (1987) provided the first treatment relating the Poisson–Dirichlet distribution to spin glasses. Successive works include Bolthausen and Sznitman (1998) about coalescence and Neveu (1992) on continuous state branching processes.

In order to present the general definition of a Poisson–Dirichlet distribution, we recall that a random variable X has the $\mathrm{Beta}(a, b)$ distribution (with $a, b > 0$) if the support is the interval $(0, 1)$ and the density is

$$f_X(x) = \frac{\Gamma(a+b)}{\Gamma(a)\Gamma(b)} x^{a-1}(1-x)^{b-1}. \tag{4.17}$$

Definition 4.5 (Griffiths–Engen–McCloskey distribution) Define a sequence $\{P_i\}_{i=1}^{\infty}$ of random variables by

$$\begin{aligned} P_1 &= V_1 \\ P_2 &= (1 - V_1)V_2 \\ P_3 &= (1 - V_1)(1 - V_2)V_3 \\ &\vdots \end{aligned} \tag{4.18}$$

[1] In the context of spin glasses it is conventional to use the letter m, whereas in other contexts the letter α is often used.

where $\{V_i\}_{i=1}^{\infty}$ are independent random variables and each V_i has distribution Beta$(1-m,\ \theta+im)$. Then the two parameter Griffiths–Engen–McCloskey distribution, GEM(m,θ), is the law of $\{P_i\}_{i=1}^{\infty}$ on the unordered infinite-dimensional simplex

$$\Delta_\infty = \left\{ \xi_1 \geqslant 0, \xi_2 \geqslant 0, \dots : \sum_{i=1}^{\infty} \xi_i = 1 \right\}. \tag{4.19}$$

Definition 4.6 (Poisson–Dirichlet distribution) Consider the ordered random sequence $\{P_{(i)}\}_{i=1}^{\infty}$, i.e. the non-increasing order statistics of the random sequence $\{P_i\}_{i=1}^{\infty}$ in Definition 4.5. Namely, $P_{(1)} \geqslant P_{(2)} \geqslant \cdots$ are the ranked values in descending order of $\{P_i\}_{i=1}^{\infty}$. Then the Poisson–Dirichlet distribution PD(m,θ) is the law of $\{P_{(i)}\}_{i=1}^{\infty}$ on $\Delta_{(\infty)}$.

The definition above has been named the *"stick-breaking construction"* because the points $\{P_i\}$ can be visualized as follows. Imagine a stick of length 1. We then break the stick into two pieces (by using a Beta-distributed random variable V_1) of lengths V_1 and $1-V_1$. P_1 is the length of the first piece. Then we repeat the breaking procedure on the second piece by using a Beta-distributed random variable V_2. Therefore we will obtain two new pieces with lengths $V_2(1-V_1) = P_2$ and $(1-V_2)(1-V_1)$. Repeatedly breaking the second piece at time n, in the limit $n \to \infty$, the whole sequence $\{P_i\}$ is constructed.

There are at least two other equivalent characterizations of the Poisson–Dirichlet distribution PD(m,θ). The first is known as the *"Chinese restaurant process"* and is obtained by the fact that a tiling of the unit interval $(0,1)$ can also be seen as a random partition of the integers $\mathbb{N}$. The process is constructed by imagining people entering a Chinese restaurant and sitting at either a new table or one which is already occupied. The process is initialized with the first person entering the restaurant and sitting at the first table. Then suppose that when the $(n+1)$th person arrives, he finds k tables which are already occupied, where table i has n_i customers ($\sum_{i=1}^{k} n_i = n$). Then person $n+1$ starts a new table with probability

$$\frac{\theta + mk}{n+\theta} \tag{4.20}$$

and sits at table i (for $1 \le i \le k$) with probability

$$\frac{n_i - m}{n+\theta}. \tag{4.21}$$

When n customers have arrived, the tables constitute the blocks of a *random partition* Π_n of $[n] = \{1,2,\dots,n\}$. Letting $n \to \infty$, these blocks have *asymptotic frequencies*, where the asymptotic frequency of a block $B \subset \mathbb{N}$ is defined to be

$$\lim_{n \to \infty} \frac{|B \cap [n]|}{n}. \tag{4.22}$$

The existence of this limit for blocks generated by the Chinese restaurant process is guaranteed by Kingman's theory of exchangeable random partitions (see Kingman (1982)). Moreover, if $(P_1, P_2, \ldots)$ is the vector of asymptotic frequencies for an (m, θ)-Chinese restaurant process in the order of the tables, then $(P_1, P_2, \ldots)$ has the GEM(m, θ) distribution. If this vector is put into decreasing order of size then it has the Poisson–Dirichlet PD(m, θ) distribution (for a proof see, for instance, Pitman (2006) or Berestycki (2009)).

A third possible approach to the Poisson–Dirichlet distribution is obtained via Poisson point processes. While the construction can be given for every (m, θ) with $\{0 \leqslant m < 1,\ \theta > -m\}$ (see Pitman (2006)), here we shall focus on the case $\theta = 0$, which is the relevant case for spin glasses. Indeed the construction of the PD$(m, 0)$ process via Poisson point processes was first considered by Ruelle in his mathematical reformulation of the Derrida random energy model (see Ruelle (1987); Derrida (1980, 1981)).

For a parameter $0 < m < 1$, consider the Poisson point process on $(0, \infty)$ that is obtained from the exponential Poisson point process on $\mathbb{R}$ applying the transformation

$$Y_i = e^{\frac{X_i}{m}} \tag{4.23}$$

to its points $\{X_i\}_{i \geqslant 1}$. Then, from Proposition 4.2, $\{Y_i\}_{i \geqslant 0}$ are the points of Poisson point process on $(0, \infty)$ with intensity $\mu(dy) = my^{-m-1}dy$. Let $\{Y_{(i)}\}_{i=1}^{\infty}$ be the ordered statistics of this Poisson point process, that is

$$Y_{(1)} > Y_{(2)} > \cdots \tag{4.24}$$

The fact that the points can be ordered is guaranteed because for all $\epsilon > 0$ in the interval (ϵ, ∞), there is almost surely a finite number of points (because $m > 0$). Moreover, one also has that the sum of the (ordered) points is almost surely finite (because $m < 1$). Thus we can consider the ordered sequence $\{P_{(1)} \geqslant P_{(2)} \geqslant \cdots\}$ where

$$P_{(i)} = \frac{Y_{(i)}}{\sum_{n=1}^{\infty} Y_{(n)}} \tag{4.25}$$

which of course belongs to $\Delta_{(\infty)}$ of Eq. (4.16). One can prove that the distribution of $\{P_{(i)}\}$ is given by the Poisson–Dirichlet distribution PD$(m, 0)$ introduced in Definition 4.6 (see, for instance, Perman *et al.* (1992)).

Much can be said about the Poisson–Dirichlet random probability distribution. For example, the following expression for the "moments" was found for PD$(m, 0)$ (see, for instance, Ruelle (1987), Derrida (1997)).

Proposition 4.7 (Identity 1 for the PD$(m, 0)$) *Let $\{P_{(i)}\}_{i\geqslant 1}$ be distributed as the Poisson–Dirichlet process* PD$(m, 0)$*, with $0 < m < 1$. The following identity holds:*

$$\mathbb{E}\left(\sum_{i\geqslant 1}(P_{(i)})^k\right) = \frac{\Gamma(k-m)}{\Gamma(1-m)\Gamma(k)}. \tag{4.26}$$

Proof We use the explicit representation of the Poisson–Dirichlet distribution in terms of the exponential Poisson point process. This yields

$$\mathbb{E}\left(\sum_{i\geqslant 1}(P_{(i)})^k\right) = \mathbb{E}\left(\sum_{i\geqslant 1}\left(\frac{e^{X_i/m}}{\sum_{j\geqslant 1} e^{X_j/m}}\right)^k\right), \tag{4.27}$$

where $\mathbb{E}$ on the righthand side denotes expectation with respect to the Poisson point process, with exponential intensity whose points are denoted by $\{X_i\}$. The denominator of (4.27) can be rewritten using the identity

$$\frac{1}{Z^k} = \frac{1}{\Gamma(k)}\int_0^{+\infty} dt\, t^{k-1}e^{-tZ}, \tag{4.28}$$

which is a direct consequence of a change of variable $tZ \to y$ and of the definition of the Gamma function

$$\Gamma(k) = \int_0^{\infty} dy\, y^{k-1}e^{-y}. \tag{4.29}$$

Thus we have

$$\mathbb{E}\left(\sum_{i\geqslant 1}(P_{(i)})^k\right) = \frac{1}{\Gamma(k)}\,\mathbb{E}\left(\sum_{i\geqslant 1} e^{X_i k/m}\int_0^{+\infty} dt\, t^{k-1}e^{-t\sum_{j=1}^{\infty} e^{X_j/m}}\right). \tag{4.30}$$

We proceed by considering the realization of the exponential Poisson point process with a cut-off c, and then send $c \to -\infty$ at the end of the computation. In the interval $(c, +\infty)$ we have a random number of points which is a Poisson random variable with parameter e^{-c}. Conditioned to having n points, these are distributed similarly to i.i.d. random variables with probability density (4.10), so the righthand side of the previous equation can be rewritten as

$$\sum_{n=0}^{\infty}\frac{1}{\Gamma(k)}\,\mathbb{E}\left(\sum_{i=1}^{n} e^{X_i k/m}\int_0^{+\infty} dt\, t^{k-1}e^{-t\sum_{j=1}^{n} e^{X_j/m}}\right)\frac{(e^{-c})^n}{n!}\,e^{-e^{-c}}, \tag{4.31}$$

Using independence of the random variables $\{X_i\}_{i=1}^n$, together with the fact that they are identically distributed, we find

$$\frac{1}{\Gamma(k)} \int_0^{+\infty} dt\, t^{k-1} \sum_{n=0}^{\infty} n\, \mathbb{E}\left(e^{X_1 k/m - t e^{X_1/m}}\right) \left(\mathbb{E}\left(e^{-t e^{X_1/m}}\right)\right)^{n-1} \frac{(e^{-c})^n}{n!} e^{-e^{-c}} \tag{4.32}$$

and using the probability density (4.10) we find

$$\frac{1}{\Gamma(k)} \int_0^{+\infty} dt\, t^{k-1} \left(\int_c^{+\infty} dx e^{xk/m - t e^{x/m} - x}\right) \sum_{n=0}^{\infty} \frac{\left(\int_c^{+\infty} dx e^{-t e^{x/m} - x}\right)^n}{n!} e^{-e^{-c}}. \tag{4.33}$$

Next, make the change of variable $e^{x/m} \to y$ and sum over n to obtain

$$\frac{1}{\Gamma(k)} \int_0^{+\infty} dt\, t^{k-1} \left(\int_{e^{c/m}}^{+\infty} dy\, m\, y^{k-m-1} e^{-ty}\right) e^{\int_{e^{c/m}}^{+\infty} dy\, m\, y^{-m-1} e^{-ty}} e^{-e^{-c}}. \tag{4.34}$$

Using the fact that

$$e^{-c} = \int_c^{+\infty} dx\, e^{-x} = \int_{e^{c/m}}^{+\infty} dy\, m\, y^{-m-1}, \tag{4.35}$$

we obtain

$$\frac{1}{\Gamma(k)} \int_0^{+\infty} dt\, t^{k-1} \left(\int_{e^{c/m}}^{+\infty} dy\, m\, y^{k-m-1} e^{-ty}\right) e^{\int_{e^{c/m}}^{+\infty} dy\, m\, y^{-m-1}(e^{-ty}-1)} \tag{4.36}$$

and making the change of variable $ty \to z$ we have

$$\frac{1}{\Gamma(k)} \int_0^{+\infty} dt\, m\, t^{m-1} \left(\int_{te^{c/m}}^{+\infty} dz\, z^{k-m-1} e^{-z}\right) e^{\int_{te^{c/m}}^{+\infty} dz\, m\, z^{-m-1}(e^{-z}-1)\, t^m}. \tag{4.37}$$

We now take the limit $c \to -\infty$. In this limit we have

$$\int_{te^{c/m}}^{+\infty} dz\, z^{k-m-1} e^{-z} \to \Gamma(k-m) \tag{4.38}$$

and, integrating by parts,

$$\int_{te^{c/m}}^{+\infty} dz\, m\, z^{-m-1}(e^{-z}-1) \to -\Gamma(1-m). \tag{4.39}$$

We therefore now arrive at

$$\frac{\Gamma(k-m)}{\Gamma(k)} \int_0^{+\infty} dt\, m\, t^{m-1} e^{-t^m \Gamma(1-m)}, \tag{4.40}$$

which can be readily integrated to obtain

$$\frac{\Gamma(k-m)}{\Gamma(1-m)\Gamma(k)}. \tag{4.41}$$

This completes the proof. □

As for the "moments", the whole "distribution" for the PD(0, m) random measures can be given.

Proposition 4.8 (Identity 2 for the PD(m, 0)) *Let* $\{P_{(i)}\}_{i\geqslant 1}$ *be distributed as the Poisson–Dirichlet process* PD(m, 0), *with* $0 < m < 1$. *For any bounded function* $f : [0,1] \to \mathbb{R}$

$$\mathbb{E}\left(\sum_{i\geqslant 1} f(P_{(i)})\right) = \frac{1}{\Gamma(m)\Gamma(1-m)} \int_0^1 du\ f(u)\, u^{-m-1}(1-u)^{m-1}. \tag{4.42}$$

Proof We consider functions $f : [0,1] \to \mathbb{R}$ which admit a series expansion

$$f(u) = \sum_{k=0}^{\infty} a_k u^k. \tag{4.43}$$

From Proposition 4.7 we have

$$\mathbb{E}\left(\sum_{i=1}^{\infty} f(P_{(i)})\right) = \mathbb{E}\left(\sum_{i=1}^{\infty}\sum_{k=0}^{\infty} a_k (P_{(i)})^k\right) = \sum_{k=0}^{\infty} a_k \frac{\Gamma(k-m)}{\Gamma(1-m)\Gamma(k)}. \tag{4.44}$$

From the definition of the Beta function

$$\mathrm{B}(x,y) = \int_0^1 du\ u^{x-1}(1-u)^{y-1}, \tag{4.45}$$

and the identity

$$\mathrm{B}(x,y) = \frac{\Gamma(x)\Gamma(y)}{\Gamma(x+y)}, \tag{4.46}$$

we obtain

$$\begin{aligned}
\mathbb{E}\left(\sum_{i=1}^{\infty} f(P_{(i)})\right) &= \sum_{k=0}^{\infty} a_k \frac{1}{\Gamma(1-m)\Gamma(m)} \mathrm{B}(k-m,m) \\
&= \sum_{k=0}^{\infty} a_k \frac{1}{\Gamma(1-m)\Gamma(m)} \int_0^1 du\ u^{k-m-1}(1-u)^{m-1} \\
&= \frac{1}{\Gamma(1-m)\Gamma(m)} \int_0^1 du\ f(u) u^{-m-1}(1-u)^{m-1}.
\end{aligned} \tag{4.47}$$

□

The computations above can be generalized to "correlation" functions by similar calculations. One obtains (Derrida (1997)) as follows.

Proposition 4.9 (Identity 3 for the PD$(m, 0)$) *Let* $\{P_{(i)}\}_{i\geqslant 1}$ *be distributed as the Poisson–Dirichlet process* PD$(m, 0)$, *with* $0 < m < 1$. *Then*

$$\mathbb{E}\left(\sum_{i=1}^{\infty}\sum_{j=1}^{\infty}(P_{(i)})^{k}(P_{(j)})^{k'}\right) = \frac{\Gamma(k+k'-m)}{\Gamma(k+k')\Gamma(1-m)} + m\frac{\Gamma(k-m)\Gamma(k'-m)}{\Gamma(k+k')[\Gamma(1-m)]^{2}}. \tag{4.48}$$

All the computations can be generalized to the generic two-parameter distribution PD(m, θ). For example the following result, which generalizes Proposition 4.8, is proved in Pitman and Yor (1997):

Proposition 4.10 (Identity 4 for the PD(m, θ)) *Let* $\{P_{(i)}\}_{i\geqslant 1}$ *be distributed as the Poisson–Dirichlet process* PD(m, θ), *with* $0 < m < 1$ *and* $\theta > -m$. *For any bounded* $f : [0, 1] \to \mathbb{R}$

$$\mathbb{E}\left(\sum_{i=1}^{\infty} f(P_{(i)})\right) = \frac{\Gamma(\theta+1)}{\Gamma(\theta+m)\Gamma(1-m)}\int_0^1 du\, f(u)\, u^{-m-1}(1-u)^{m+\theta-1}. \tag{4.49}$$

4.4 The random energy model

The random energy model (REM) is a statistical mechanical model, where the energy levels are independent and identically distributed Gaussian random variables. For a system of size N, the Hamiltonian is a Gaussian-centered family defined by its covariance matrix

$$\mathcal{C}_N(\sigma, \sigma') := \mathbb{E}[H_N(\sigma)H_N(\sigma')] = \frac{N}{2}\,\delta(\sigma, \sigma') \tag{4.50}$$

where $\sigma, \sigma' \in \Sigma_N = \{+1, -1\}^N$ are vectors, whose components are Ising spin variables. The dependence on N in (4.50) is such that thermodynamic quantities (internal energy, free energy, etc.) are extensive in the volume, while the factor $1/2$ is included as a matter of convention.

In terms of the general Hamiltonian representation introduced in Definition 1.4, the REM Hamiltonian is given by

$$H_N(\sigma) = \sum_{X\subset\{1,2,\ldots,N\}} J_X\sigma_X \tag{4.51}$$

with J_X a set of centered independent Gaussian with variance $\mathbb{E}(J_X^2) = N/2^{N+1}$. This can be readily seen using the representation Theorem 2.3. From this point, we will use the more immediate representation

$$H_N(\sigma) = \sqrt{\frac{N}{2}}\, X(\sigma) \tag{4.52}$$

where $\{X(\sigma)\}_{\sigma\in\Sigma_N}$ are 2^N i.i.d. copies of a standard normal random variable.

The model was introduced and solved in Derrida (1980, 1981). A mathematically rigorous treatment can be found in the books by Bovier (2006) and Talagrand (2010b). There are other mathematical analyses, like Olivieri and Picco (1984), and also treatments which rely more on large-deviation theory, such as Dorlas and Wedagedera (2001), and Eisele (1983). See also Bovier and Kurkova (2004a, b).

In this section, following Giardinà and Starr (2007), we will compute the pressure using the variational approach that has been developed for the SK model. To this end, we adapt the quadratic interpolation method to obtain an asymptotically sharp upper bound on the pressure (this is the analog, for the REM, of Guerra's bounds for the SK model that will be discussed in Section 4.12). Then we show that the upper bound is optimal by proving a corresponding lower bound. This is based on a simple induction argument which gives full control of the high temperature region, and then on a straightforward application of convexity of the pressure.

We recall the standard definitions. The main quantity we are going to study through variational principles is the quenched *pressure*

$$p_N(\beta) = \frac{1}{N}\,\mathbb{E}[\ln \mathcal{Z}_N(\beta)]. \tag{4.53}$$

We will be interested in the thermodynamic limit

$$p(\beta) = \lim_{N\to\infty} p_N(\beta). \tag{4.54}$$

The result is as follows.

Theorem 4.11 (Pressure REM) *With the inverse critical temperature defined as $\beta_c = 2\sqrt{\ln 2}$, one has*

$$p(\beta) = \bar{p}(\beta) := \begin{cases} \frac{1}{4}(\beta^2 + \beta_c^2) & \text{for } \beta < \beta_c \\ \frac{1}{2}\beta\beta_c & \text{for } \beta \geq \beta_c. \end{cases} \tag{4.55}$$

Proof The theorem is proved by a combination of an upper and lower bound which are shown, respectively, in Sections 4.5 and 4.6. □

4.5 Upper bound for the REM pressure

We start by recalling the quadratic interpolation technique, first identified in Guerra and Toninelli (2002), and further developed in Aizenman *et al.* (2003), which has led to many advances in the mathematical study of spin glasses.

Proposition 4.12 (Pressure upper bound) *Let $H(\sigma)$ be a Gaussian family, indexed by $\sigma \in \Sigma_N$, with covariance $C^{(H)}(\sigma;\sigma')$. Let $\alpha \in \mathcal{A}$ be an index ranging over the set $\mathcal{A}$, and let $K(\alpha)$ and $V(\sigma,\alpha)$ be Gaussian random variables, independent of $H(\sigma)$ and of each other, with covariances $C^{(K)}(\alpha;\alpha')$ and $C^{(V)}(\sigma,\alpha;\sigma',\alpha')$, respectively. Suppose that*

$$C^{(H)}(\sigma;\sigma') + C^{(K)}(\alpha;\alpha') \geqslant C^{(V)}(\sigma,\alpha;\sigma',\alpha') \tag{4.56}$$

for all $\sigma,\sigma' \in \Sigma_N$ and $\alpha,\alpha' \in \mathcal{A}$, and suppose that

$$C^{(H)}(\sigma;\sigma) + C^{(K)}(\alpha;\alpha) = C^{(V)}(\sigma,\alpha;\sigma,\alpha) \tag{4.57}$$

for every $\sigma \in \Sigma_N$ and $\alpha \in \mathcal{A}$. Moreover, suppose that there is a random weight $w : \mathcal{A} \to [0,\infty)$ such that, almost surely, $\sum_{\alpha\in\mathcal{A}} w(\alpha)$ is strictly positive and finite. Then,

$$p_N(\beta) \leqslant \mathbb{E}\left[\frac{1}{N}\ln\frac{\sum_{\sigma,\alpha} w(\alpha)e^{-\beta V(\sigma,\alpha)}}{\sum_\alpha w(\alpha)e^{-\beta K(\alpha)}}\right]. \tag{4.58}$$

Proof We follow the interpolation method generally introduced in Theorem 3.46. For $t \in [0,1]$, define an interpolating Hamiltonian

$$\tilde{H}(\sigma,\alpha;t) = \sqrt{1-t}\,[H(\sigma)+K(\alpha)] + \sqrt{t}\,V(\sigma,\alpha). \tag{4.59}$$

and an associated random partition function

$$\mathcal{Z}_{N,t}(\beta) = \sum_{\sigma,\alpha} w(\alpha)e^{-\beta\tilde{H}(\sigma,\alpha;t)}. \tag{4.60}$$

Let $\Omega_{N,\beta,t}$ denote expectation with respect to the multiple replica product measure, where the weight for a configuration (σ,α) of a generic copy is given by Gibbs measure associated with $\tilde{H}(\sigma,\alpha;t)$ times a generic weight $w(\alpha)$. In particular, for a function $f(\sigma,\alpha,\sigma',\alpha')$ of two replicas, one has

$$\Omega_{N,\beta,t}\{f\} = \sum_{\sigma,\alpha}\sum_{\sigma',\alpha'} w(\alpha)w(\alpha')\,\frac{e^{-\beta[\tilde{H}(\sigma,\alpha;t)+\tilde{H}(\sigma',\alpha';t)]}}{\mathcal{Z}^2_{N,t}}\,f(\sigma,\alpha;\sigma',\alpha'). \tag{4.61}$$

Then one has (because of the equality along the diagonal Eq. (4.57))

$$\frac{1}{N}\,\mathbb{E}\Big[\ln\sum_{\sigma,\alpha} w(\alpha)e^{-\beta V(\sigma,\alpha)}\Big] - \frac{1}{N}\,\mathbb{E}\Big[\ln\sum_{\alpha} w(\alpha)e^{-\beta K(\alpha)}\Big] - p_N(\beta)$$
$$= \frac{\beta^2}{2}\int_0^1 \mathbb{E}[\Omega_{N,\beta,t}\{C^{(H)} + C^{(K)} - C^{(V)}\}]\,dt. \tag{4.62}$$

This is proved by differentiating the quantity

$$\mathbb{E}\left[N^{-1}\ln\sum\nolimits_{\alpha,\sigma} w(\alpha)e^{-\beta\tilde{H}_N(\sigma,\alpha;t)}\right], \tag{4.63}$$

with respect to t and using integration by parts. Because of (4.56), the righthand side of (4.62) is obviously positive and Eq. (4.58) follows. □

Remark 4.13 An identity such as (4.62) is usually called a sum rule. The process $K(\alpha)$ has to be thought of as a large reservoir which acts on the original system $H(\sigma)$ through the interaction $V(\sigma,\alpha)$. We will return to this when discussing the extended variational principle for the SK model in Section 4.11.

We are going to use the previous lemma to establish an optimal upper bound for the REM. A key element is to choose the correct formula for the random weight $w(\alpha)$. The right choice for mean-field spin glasses seems to be generally given by Ruelle's random probability cascade. For the REM it is given by a single level of the cascade, yielding (when normalized) the Poisson–Dirichlet process PD(m, 0).

In order to apply the previous proposition to the REM, the invariance property of the exponential Poisson point process $\{Y_\alpha\}$ under the addition of an independent sequence of random variables $\{F_\alpha\}$ described in Proposition 4.3 will be useful. Namely,

$$\sum_{\alpha=1}^{\infty} e^{Y_\alpha/m}\exp(F_\alpha) \overset{\mathcal{D}}{=} \mathbb{E}[\exp(mF)]^{1/m}\sum_{\alpha=1}^{\infty} e^{Y_\alpha/m}. \tag{4.64}$$

Also, we will need an appropriate choice of the auxiliary processes K and V; this is the content of the following.

Proposition 4.14 (Upper bound REM) *Let $\mathcal{C}^{(H)}(\sigma,\sigma')$ be the covariance of the REM, given by Eq. (4.50). Choose $w(\alpha)=\exp[y(\alpha)/m]$ for $0<m<1$. For each $b\geqslant 1$, let*

$$\mathcal{C}^{(K)}(\alpha,\alpha') = (b-1)\frac{N}{2}\,\delta(\alpha,\alpha') \tag{4.65}$$

$$\mathcal{C}^{(V)}(\sigma,\alpha;\sigma',\alpha') = b\,\frac{N}{2}\,\delta(\sigma,\sigma')\delta(\alpha,\alpha'). \tag{4.66}$$

Then one has the upper bound for the REM,

$$p_N(\beta) \leqslant \inf_{0<m<1}\left[\frac{1}{4}m\beta^2 + \frac{1}{m}\ln 2\right]. \tag{4.67}$$

If we compute the infimum, we obtain

$$p_N(\beta) \leqslant \bar{p}(\beta) \tag{4.68}$$

with $\beta_c = 2\sqrt{\ln 2}$ *and*

$$\bar{p}(\beta) = \begin{cases} \frac{1}{4}(\beta^2 + \beta_c^2) & \textit{for } \beta < \beta_c \\ \frac{1}{2}\beta\beta_c & \textit{for } \beta \geq \beta_c. \end{cases} \tag{4.69}$$

Proof We note that Proposition 4.12 is applicable because, by inspection,

$$b\,\delta(\sigma, \sigma')\delta(\alpha, \alpha') \leqslant \delta(\sigma, \sigma') + (b-1)\delta(\alpha, \alpha') \tag{4.70}$$

From Eq. (4.66), we have

$$\begin{aligned} &\frac{1}{N}\mathbb{E}\left[\ln \sum_{\sigma,\,\alpha} w(\alpha) \exp[-\beta V(\sigma, \alpha)]\right] \\ &= \frac{1}{N}\mathbb{E}\left[\ln \sum_{\alpha} \exp\left[\frac{Y_\alpha}{m}\right] \sum_{\sigma} \exp\left[-\beta\sqrt{\frac{bN}{2}} X(\sigma, \alpha)\right]\right] \\ &= \frac{1}{N}\mathbb{E}\left[\ln \sum_{\alpha} \exp\left[\frac{Y_\alpha}{m}\right] \mathcal{Z}_N(\beta\sqrt{b}; \alpha)\right] \end{aligned} \tag{4.71}$$

where $\mathcal{Z}_N(\beta\sqrt{b}; \alpha)$ are independent copies (labeled by the αs) of the random variable $\mathcal{Z}_N(\beta\sqrt{b}) = \sum_\sigma \exp[-\beta\sqrt{\frac{bN}{2}} X(\sigma)]$. By applying the invariance property of Eq. (4.64) to Eq. (4.71) with $\exp[F_\alpha] = \mathcal{Z}_N(\beta\sqrt{b}; \alpha)$, we obtain

$$\frac{1}{N}\mathbb{E}\left[\ln \sum_{\sigma,\,\alpha} w(\alpha) \exp[-\beta V(\sigma, \alpha)]\right] \tag{4.72}$$

$$= \frac{1}{N}\mathbb{E}\left[\ln \sum_{\alpha} \exp\left[\frac{Y_\alpha}{m}\right]\right] + \frac{1}{mN}\ln \mathbb{E}\left[\mathcal{Z}_N^m(\beta\sqrt{b})\right]. \tag{4.73}$$

Analogously, due to the choice (4.65), we have

$$\begin{aligned} &\frac{1}{N}\mathbb{E}\left[\ln \sum_{\alpha} \exp\left[\frac{Y_\alpha}{m}\right] \exp[-\beta K(\alpha)]\right] \\ &= \frac{1}{N}\mathbb{E}\left[\ln \sum_{\alpha} \exp\left[\frac{Y_\alpha}{m}\right] \exp\left[-\beta\sqrt{\frac{(b-1)N}{2}} X_\alpha\right]\right]. \end{aligned} \tag{4.74}$$

Using the invariance property Eq. (4.64) again, with

$$\exp[F_\alpha] = \exp\left[-\beta\sqrt{\frac{(b-1)N}{2}} X_\alpha\right],$$

we obtain

$$\frac{1}{N}\mathbb{E}\left[\ln\sum_{\alpha}\exp\left[\frac{Y_{\alpha}}{m}\right]\exp[-\beta K(\alpha)]\right] = \tag{4.75}$$

$$\frac{1}{N}\mathbb{E}\left[\ln\sum_{\alpha}\exp\left[\frac{Y_{\alpha}}{m}\right]\right] + \beta^2\frac{(b-1)m}{4}. \tag{4.76}$$

Putting together Eqs (4.73) and (4.76) we obtain:

$$p_N(\beta) \leqslant \frac{1}{mN}\ln\mathbb{E}\left[\mathcal{Z}_N^m(\beta\sqrt{b})\right] - \beta^2\frac{(b-1)m}{4}. \tag{4.77}$$

Now we use the simple fact that

$$\mathcal{Z}_N^m(\beta\sqrt{b}) \leq \mathcal{Z}_N(m\beta\sqrt{b}). \tag{4.78}$$

This is a general fact in statistical mechanics: since the entropy is positive by definition, the free energy is increasing in β. Indeed, considering $f_N(\beta) = -\frac{1}{N\beta}\ln(\mathcal{Z}_N(\beta))$, the *random* free energy, one immediately checks that $f_N'(\beta) = \frac{1}{\beta}\left(u_N(\beta) - f_N(\beta)\right) = \frac{1}{\beta^2}s_N(\beta) \geq 0$, where $u_N(\beta)$ is the *random* internal energy, and $s_N(\beta)$ is the *random* entropy. Therefore, for any $0 < m \leq 1$, we have:

$$-\frac{1}{Nm\beta}\ln(\mathcal{Z}_N(m\beta)) \leq -\frac{1}{N\beta}\ln(\mathcal{Z}_N(\beta)), \tag{4.79}$$

which is equivalent to (4.78) when we replace β by $\beta\sqrt{b}$. By inserting Eq. (4.78) into Eq. (4.77), it is now easy to compute the expectation

$$\mathbb{E}\left[\mathcal{Z}_N(m\beta\sqrt{b})\right]$$

and we arrive at the upper bound

$$p_N(\beta) \leqslant \frac{1}{4}m\beta^2 + \frac{1}{m}\ln 2. \tag{4.80}$$

Note that there is no longer any dependence on b in the bound. Finally, the optimal bound is obtained by minimizing m which yields Eq. (4.67). The inequality of (4.67) takes two different forms depending on whether β is greater or less than $\beta_c := 2\sqrt{\ln(2)}$. For $\beta > \beta_c$, the righthand side of (4.67) is optimized at $m = \beta_c/\beta$. For $\beta < \beta_c$, the infimum over $0 < m < 1$ is attained by a limit $m \to 1$. The upper bound in Eq. (4.69) is deduced and this completes the proof. □

Remark 4.15 We have introduced the use of the extended variational principle to obtain an upper bound on the REM pressure for pedagogical reasons, to pave the way toward the analogous treatment of the SK model. For the REM we notice, however, that the pressure upper bound can be viewed as an instance of Jensen's inequality:

$$\mathbb{E}[\ln(\mathcal{Z}_N(\beta))] = \frac{1}{m}\,\mathbb{E}[\ln \mathcal{Z}_N^m(\beta)] \leq \frac{1}{m} \ln \mathbb{E}[\mathcal{Z}_N^m(\beta)]. \tag{4.81}$$

This holds for $m > 0$, but one needs $m \leq 1$ to apply (4.78).

4.6 Lower bound for the REM pressure

Since, by Proposition 4.14, $p_N(\beta) \leq \bar{p}(\beta)$ for all N, it follows that $p(\beta) \leq \bar{p}(\beta)$ in the limit $N \to \infty$. We want to show that the opposite is also true: to establish that $p(\beta) = \bar{p}(\beta)$ for all $\beta \geq 0$. The key to obtaining the lower bound is to understand the high temperature region. Here we follow Giardinà and Starr (2007); see also Talagrand (2010b), Proposition 1.1.5; and Bovier (2006), Theorem 9.1.2, where the "truncated second moment method" is used.

Proposition 4.16 (Lower bound REM) *For $\beta \in [0, +\infty)$*

$$p(\beta) \geqslant \bar{p}(\beta) \tag{4.82}$$

with

$$\bar{p}(\beta) = \begin{cases} \frac{1}{4}(\beta^2 + \beta_c^2) & \text{for } \beta < \beta_c \\ \frac{1}{2}\beta\beta_c & \text{for } \beta \geq \beta_c. \end{cases} \tag{4.83}$$

Proof The proof of the proposition is made in two separate ways according to the high ($\beta < \beta_c$) or low ($\beta > \beta_c$) temperature range. The high temperature region is proved in Lemma 4.17 and the low temperature in Lemma 4.21. □

Lemma 4.17 (High temperature)

$$p(\beta) = \frac{\beta^2}{4} + \log(2) = \bar{p}(\beta) \quad \text{for } \beta \leq \beta_c. \tag{4.84}$$

Proof A straightforward computation gives

$$p_N(\beta) = \ln(2) + \frac{\beta^2}{4} - \int_0^\beta \frac{\beta'}{2}\,\mathbb{E}[\Omega_{N,\beta'}\{\delta(\sigma,\sigma')\}]d\beta' \tag{4.85}$$

from which the lemma follows, once we prove that the integrand converges to zero in the large-volume limit. We will prove that such a convergence is exponentially fast in the volume (uniformly in the temperature) in Lemma 4.20 (which makes use of Lemmata 4.18 and 4.19). □

Lemma 4.18 *Let $\Omega_{\beta,N}$ refer to the (expectation associated to the) random probability measure on $\Sigma_N \times \Sigma_N$ specified by*

$$\Omega_{\beta,N}\{f(\sigma,\sigma')\} := \mathcal{Z}_N^{-2}(\beta)\sum\nolimits_{\sigma,\sigma'\in\Sigma_N} e^{-\beta H(\sigma)}e^{-\beta H(\sigma')}f(\sigma,\sigma'). \quad (4.86)$$

For $0 \le \beta \le \beta_c$ we have

$$\frac{1}{N}\ln \mathbb{E}[\mathcal{Z}_N(\beta)\Omega_{\beta,N}\{\delta(\sigma,\sigma')\}] \le \frac{\beta\beta_c}{2}. \quad (4.87)$$

Proof Note that

$$\Omega_{\beta,N}\{\delta(\sigma,\sigma')\} = \frac{\sum_{\sigma,\sigma'\in\Sigma_N} e^{-\beta H(\sigma)}e^{-\beta H(\sigma')}\delta(\sigma,\sigma')}{\mathcal{Z}_N^2(\beta)} = \frac{\mathcal{Z}_N(2\beta)}{\mathcal{Z}_N^2(\beta)}. \quad (4.88)$$

By (4.78), we know that for $0 < m < 1$

$$\mathcal{Z}_N(2\beta) \le \mathcal{Z}_N^{1/m}(2m\beta). \quad (4.89)$$

Using Hölder's inequality with $p = 1/m$, $q = 1/(1-m)$ and $\frac{1}{p}+\frac{1}{q} = 1$

$$\mathcal{Z}_N(2m\beta) \le \mathcal{Z}_N^m(\beta)\,\mathcal{Z}_N^{1-m}\Big(\frac{m}{1-m}\beta\Big). \quad (4.90)$$

Therefore,

$$\mathcal{Z}_N(\beta)\Omega_{\beta,N}\{\delta(\sigma,\sigma')\} = \frac{\mathcal{Z}_N(2\beta)}{\mathcal{Z}_N(\beta)} \le \frac{\mathcal{Z}_N^{1/m}(2m\beta)}{\mathcal{Z}_N(\beta)} \le \mathcal{Z}_N^{(1-m)/m}\Big(\frac{m}{1-m}\beta\Big).$$

Since $m/(1-m)$ can take any positive value as m ranges over $(0,1)$, this means that

$$\mathcal{Z}_N(\beta)\Omega_{\beta,N}\{\delta(\sigma,\sigma')\} \le \mathcal{Z}_N^{1/r}(r\beta) \quad (4.91)$$

for every $r \in [0,+\infty)$. Moreover, for $r \ge 1$ we can use Jensen's inequality to obtain

$$\mathbb{E}[\mathcal{Z}_N^{1/r}(r\beta)] \le (\mathbb{E}[\mathcal{Z}_N(r\beta)])^{1/r} = \exp\Big(N\Big[\frac{\beta_c^2}{4r}+\frac{r\beta^2}{4}\Big]\Big). \quad (4.92)$$

It is easy to see that the optimal value $r = \beta_c/\beta$, which satisfies the constraint $r \ge 1$ because of the hypothesis $\beta \le \beta_c$. Choosing this r and combining (4.91) and (4.92) yields (4.87). □

A second useful estimate is the following formula which gives the general concentration of measure property of Theorem 3.10, specializing in the REM.

Lemma 4.19 *For any β, and any $t > 0$,*

$$\mathbb{P}\{|N^{-1}\ln \mathcal{Z}_N(\beta) - p_N(\beta)| \ge \beta t\} \le 2e^{-Nt^2/2}.$$

Finally, the following lemma proves the exponential convergence to zero of the order parameter in the high temperature phase.

Lemma 4.20 *For any* $0 \leq \beta < \beta_c$,

$$\limsup_{N\to\infty} \sup_{0\leq\beta'\leq\beta} \frac{1}{N} \ln \mathbb{E}[\Omega_{\beta',N}\{\delta(\sigma,\sigma')\}] < 0. \tag{4.93}$$

Proof The proof will obtained by induction. Let us define the succession of temperatures given by $\beta_0 = 0$, $\beta_{n+1} = g(\beta_n)$ for $n \in \mathbb{N}$, where g is a given function. As we will see, we can choose

$$g(\beta) = \beta + a\beta_c[1 - (\beta/\beta_c)]^2 \quad \text{for any } 0 < a < 1/2. \tag{4.94}$$

and it will follow that $\beta_n \uparrow \beta_c$ as $n \to \infty$.

We first note that (4.93) is true for $\beta = \beta_0 = 0$, because one has

$$\mathbb{E}[\Omega_{0,N}\{\delta(\sigma,\sigma')\}] = 2^{-N}. \tag{4.95}$$

For the induction step, we will prove that if (4.93) is true for $\beta \in [0, \beta_n]$, then it is also true for every $\beta \in [0, \beta_{n+1}]$. Then, since $\beta_n \uparrow \beta_c$ the statement of Lemma 4.20 follows.

Suppose now that (4.93) is true for $\beta \in [0, \beta_n]$. Consider a generic $\beta \in (\beta_n, \beta_{n+1}]$ and for $t > 0$ let $A_N(\beta, t)$ be the event in Lemma 4.19:

$$A_N(\beta, t) = \{|N^{-1} \ln \mathcal{Z}_N(\beta) - p_N(\beta)| \geq \beta t\}. \tag{4.96}$$

On $A_N(\beta, t)^c$ we have $\mathcal{Z}_N(\beta) \geq e^{N[p_N(\beta)-\beta t]}$. We can now conclude that:

$$\begin{aligned}\mathbb{E}[\Omega_{\beta,N}\{\delta(\sigma,\sigma')\}\mathbb{1}_{\{A_N(\beta,t)^c\}}] &\leq \mathbb{E}\Big[\frac{\mathcal{Z}_N(\beta)}{e^{N[p_N(\beta)-\beta t]}}\Omega_{\beta,N}\{\delta(\sigma,\sigma')\}\mathbb{1}_{\{A_N(\beta,t)^c\}}\Big] \\ &\leq e^{-N[p_N(\beta)-\beta t]}\mathbb{E}[\mathcal{Z}_N(\beta)\Omega_{N,\beta}\{\delta(\sigma,\sigma')\}]. \end{aligned}\tag{4.97}$$

Therefore

$$\frac{1}{N} \ln \mathbb{E}[\Omega_{N,\beta}\{\delta(\sigma,\sigma')\}\mathbb{1}_{\{A_N(\beta,t)^c\}}] \leq -p_N(\beta) + \frac{1}{2}\beta\beta_c + \beta t \tag{4.98}$$

follows from Lemma 4.18.

Since we assumed $\beta > \beta_n$, a lower bound for $p_N(\beta)$ is given by $p_N(\beta_n)$ and a lower bound for $p_N(\beta_n)$ is given by $\frac{1}{4}[\beta_c^2 + \beta_n^2] - o(1)$, using the induction hypothesis, where $o(1)$ represents a quantity whose limit is 0 when $N \to \infty$. Therefore,

$$\frac{1}{N} \ln \mathbb{E}[\Omega_{\beta,N}\{\delta(\sigma,\sigma')\}, A_N(\beta,t)^c] \leq -\frac{1}{4}(\beta_c - \beta_n)^2 + \frac{1}{2}\beta_c[\beta - \beta_n] + \beta t - o(1). \tag{4.99}$$

On the other hand, one always has $0 \leq \Omega_{\beta,N}\{\delta(\sigma,\sigma')\} \leq 1$. Hence,

$$\mathbb{E}[\Omega_{\beta,N}\{\delta(\sigma,\sigma')\}1\!\!1_{\{A_N(\beta,t)\}}] \leq \mathbb{E}[1\!\!1_{\{A_N(\beta,t)\}}] = \mathbb{P}(A_N(\beta,t)).$$

So, by Lemma 4.19,

$$\frac{1}{N}\ln \mathbb{E}[\Omega_{\beta,N}\{\delta(\sigma,\sigma')\}1\!\!1_{\{A_N(\beta,t)\}}] \leq -\frac{1}{2}t^2 + N^{-1}\ln(2). \tag{4.100}$$

Combining Eqs (4.99) and (4.100), we obtain

$$\frac{1}{N}\ln \mathbb{E}[\Omega_{\beta,N}\{\delta(\sigma,\sigma')\}] \leq N^{-1}\ln(2) + \max\left\{-\frac{1}{2}t^2 + N^{-1}\ln(2),\ -\frac{1}{4}(\beta_c-\beta_n)^2 + \frac{1}{2}\beta_c[\beta-\beta_n] + \beta t - o(1)\right\}. \tag{4.101}$$

If we now take

$$\beta < \beta_n + \frac{1}{2}\beta_c[1-(\beta_n/\beta_c)]^2$$

then it is clear that by choosing a small enough positive t, we will have a strictly negative limsup on the lefthand side of (4.101) as $N \to \infty$. Choosing any $0 < a < 1/2$, let us take $g(\beta) = \beta + a\beta_c[1-(\beta/\beta_c)]^2$. Then for $\beta_{n+1} = g(\beta_n)$, we have proved the induction step: for β in the range $[0, \beta_{n+1}]$, inequality (4.93) also holds. □

The previous lemmata conclude the high temperature proof. The result in the low temperature phase follows as a consequence of convexity.

Lemma 4.21 (Low temperature)

$$p(\beta) \geqslant \bar{p}(\beta) \quad \textit{for } \beta \geqslant \beta_c. \tag{4.102}$$

Proof It is a basic fact, easily seen from Definition (4.53), that $p_N(\beta)$ is convex in β for all N. Therefore, the limiting function $p(\beta)$ is also convex. Hence, for any $\beta \geq \beta_c$, we have

$$p(\beta) \geqslant p(\beta_c) + (\beta-\beta_c)p'(\beta_c^-), \tag{4.103}$$

where $p'(\beta_c^-)$ is the left derivative in β_c. Since $p(\beta_c) = \bar{p}(\beta_c) = 2\ln(2)$ and $p'(\beta_c^-) = \bar{p}'(\beta_c^-) = \sqrt{\ln(2)}$, the previous formula gives $p(\beta) \geqslant \bar{p}(\beta)$ for $\beta \geqslant \beta_c$. □

4.7 Statistics of energy levels for the REM

There is much interest in the statistics of energy levels in the REM because in the bulk there is a kind of universality. Recently, Bauke and Mertens (2004) conjectured that the local statistics of energies in random spin systems with discrete spin space should, in most circumstances, be the same as in the REM. Bovier and Kurkova (2006) gave necessary conditions for this hypothesis to be true, which they showed to be satisfied in a wide range of examples, including short-range spin glasses and mean-field spin glasses of the SK type.

In the low temperature region, the statistics of the energy levels of the REM, after a suitable rescaling, are conveniently described by the exponential Poisson point process.

Theorem 4.22 *Let $\sigma^{(i)}$, for $i = 1, \ldots, 2^N$, enumerate the configurations of a spin system of size N and $H_N(\sigma^{(i)})$ be the Hamiltonian of the REM. Then the point process whose points are defined by*

$$X_{N,i} = \begin{cases} 2\sqrt{\ln 2}[H_N(\sigma^{(i)}) - b_N] & \text{if } i \leqslant 2^N \\ 0 & \text{otherwise} \end{cases}$$

with

$$b_N^2 = N^2 \ln 2 - N \ln \sqrt{4\pi N \ln 2}, \tag{4.104}$$

converges, in the limit $N \to \infty$, to the Poisson point process with intensity measure $e^{-x}dx$.

Proof We observe that for $i \leqslant 2^N$, the $X_{N,i}$s are Gaussian with mean $\mathbb{E}(X_{N,i}) = -2\sqrt{\ln 2}\, b_N$ and variance $\mathrm{Var}(X_{N,i}) = 4 \ln 2 \mathrm{Var}(H_N(i)) = 2N \ln 2$. Therefore they have a density function given by

$$f_{X_{N,i}}(x) = \frac{1}{\sqrt{4\pi N \ln 2}} e^{-\frac{(x+2\sqrt{\ln 2} b_N)^2}{4N \ln 2}}. \tag{4.105}$$

Then we compute the moment-generating function of the random variable $M_N(A)$ which counts the number of points of the process $X_{N,i}$ which belong to the interval $A = (a, b)$. We have

$$\begin{aligned} \mathbb{E}(e^{-tM_N(A)}) &= \mathbb{E}\left(e^{-t\sum_{i=1}^{2^N} \mathbb{1}_{X_{N,i}}(A)}\right) \\ &= \left(\mathbb{E}(e^{-t\mathbb{1}_{X_{N,1}}(A)})\right)^{2^N} \\ &= \left(1 + (e^{-t} - 1)\mu_N(A)\right)^{2^N} \end{aligned} \tag{4.106}$$

where

$$\mu_N(A) = \int_A f_{X_{N,1}}(x)dx. \quad (4.107)$$

If we can find a sequence b_N such that

$$\lim_{N\to\infty} 2^N \mu_N(A) = \mu(A) = \int_A e^{-x}dx \quad (4.108)$$

then we obtain

$$\lim_{N\to\infty} \mathbb{E}(e^{-tM_N(A)}) = e^{(e^{-t}-1)\mu(A)} \quad (4.109)$$

thus implying that $M(A) = \lim_{N\to\infty} M_N(A)$ is a Poisson random variable with parameter $\mu(A)$. One immediately checks from Eq. (4.105) that such a sequence b_N is indeed given by (4.104). The verification of the asymptotic independence properties of $M_N(A)$ and $M_N(B)$ when A and B are disjoint sets is left to the reader. All in all, we have that the sequence $\{X_{N,i}\}$ converges for $N \to \infty$ to the points of the Poisson point process with exponential intensity. □

4.8 The generalized random energy model

In order to treat the generalized random energy model (GREM), we extend the method developed in the previous section. The GREM is essentially a "correlated random energy model" on a hierarchical graph, i.e. a tree. The model has been introduced and solved in Derrida (1985); Derrida and Gardner (1986). Here we will follow the formulation of the GREM given in Contucci *et al.* (2003).

The GREM is a family of models, taking various parameters for their definition. Let $n \in \mathbb{N}_+$ be an integer equal to the number of levels in the hierarchical tree. Let $K_1, \dots, K_n$ be positive integers such that $K_1 + K_2 + \cdots + K_n = N$, where N is the system size. Also, let $a_1, \dots, a_n$ be real numbers such that $0 < a_i$ for $i = 1, \dots, n$ and $a_1 + a_2 + \cdots + a_n = 1$.

Definition 4.23 (GREM) Given $\sigma \in \Sigma_N$, for $i = 1, \dots, n$, let $\pi_i(\sigma)$ be the canonical projection over the subset Σ_{K_i} generated by the lexicographical partition $\mathcal{P}$ of the coordinates $(\sigma_1, \dots, \sigma_N)$ into the first K_1 coordinates, the successive K_2 coordinates and so on, up to the last K_n coordinates. Namely, $\Sigma_N = \Sigma_{K_1} \times \cdots \times \Sigma_{K_n}$, $\otimes_{i=1}^n \pi_i = 1_{\Sigma_N}$ and $\pi_i(\sigma) = (\sigma_{K_1+\cdots+K_{i-1}+1}, \dots, \sigma_{K_1+\cdots+K_i})$.

Then the GREM Hamiltonian is a family of Gaussian random variables having the covariance

$$\mathbb{E}\left[H_N(\sigma)H_N(\sigma')\right] = \frac{N}{2}\sum_{i=1}^{n} a_i \prod_{j=1}^{i} \delta(\pi_j(\sigma), \pi_j(\sigma')). \tag{4.110}$$

An explicit form of GREM Hamiltonian is

$$H_N(\sigma) = \sqrt{\frac{N}{2}}\sum_{i=1}^{n} \sqrt{a_i}\, X(\pi_1(\sigma), \ldots, \pi_i(\sigma)) \tag{4.111}$$

where, for each $i = 1, \ldots, n$, the family of random variables

$$\{X(\pi_1(\sigma), \ldots, \pi_i(\sigma))\}_{\sigma\in\Sigma_N}$$

is $2^{K_1+K_2+\cdots+K_i}$ i.i.d. Gaussians, and each family is independent of the others.

Remark 4.24 The Hamiltonian (4.111) corresponds to a tree with a branching number that at each level is a power of two. We will stick to this case to simplify the notation, while the more general case of arbitrary branching numbers (with the constraint of having approximately 2^N leaves in the last layer) is completely equivalent in the thermodynamic limit. In order to make statements that apply in the limit, we will consider sequences of Ns and $K_1, \ldots, K_n$ such that there are rational numbers $\kappa_1, \ldots, \kappa_n$, all non-negative and summing to 1, with $K_i = N\kappa_i$ for each $i = 1, \ldots, n$.

We now prove the variational expression for the pressure of the GREM. The strategy is to apply the results obtained for the REM in the previous section at each level in the hierarchy. In order to denote the dependence on the parameters $a = (a_1, \ldots, a_n)$ and $\kappa = (\kappa_1, \ldots, \kappa_n)$, let us write the GREM pressure as

$$p_N^{(n)}(\beta; a, \kappa) = \frac{1}{N}\mathbb{E}[\ln \mathcal{Z}_N(\beta; a, \kappa)] \tag{4.112}$$

and its thermodynamic limit as $p^{(n)}(\beta; a, \kappa) := \lim_{N\to\infty} p_N^{(n)}(\beta; a, \kappa)$.

4.9 Upper bound for the GREM pressure

Proposition 4.25 *Consider the GREM, for which $C^{(H)}(\sigma, \sigma')$ is given by (4.110). For an index $\alpha = (\alpha_1, \alpha_2, \ldots, \alpha_n) \in \mathcal{A}^n$, let the random weights $w(\alpha)$ be given by*

$$w(\alpha_1, \ldots, \alpha_n) = \exp\left[\frac{y(\alpha_1)}{m_1}\right]\exp\left[\frac{y(\alpha_1, \alpha_2)}{m_2}\right]\cdots\exp\left[\frac{y(\alpha_1, \ldots, \alpha_n)}{m_n}\right] \tag{4.113}$$

where the Poisson point processes are now a cascade with intensity measure $e^{-y}dy$. Namely, $y(\alpha_1)$ is the usual Poisson point process, then for each given α_1, $y(\alpha_1, \alpha_2)$ is an independent copy (labeled by α_1) of the Poisson point process, ... and so on up to $y(\alpha_1, \dots, \alpha_n)$ which, for each given $\alpha_1, \dots, \alpha_{n-1}$, is an independent copy of the Poisson point process labeled by $\alpha_1, \dots, \alpha_{n-1}$. We also choose a sequence $0 < m_1 \le m_2 \le \dots \le m_n < 1$ and

$$\mathcal{C}^{(K)}(\alpha, \alpha') = (b-1)\frac{N}{2}\sum_{i=1}^{n} a_i \prod_{j=1}^{i} \delta(\alpha_j, \alpha'_j) \tag{4.114}$$

$$\mathcal{C}^{(V)}(\sigma, \alpha, \sigma', \alpha') = b\frac{N}{2}\sum_{i=1}^{n} a_i \prod_{j=1}^{i} \delta(\pi_j(\sigma), \pi_j(\sigma'))\delta(\alpha_j, \alpha'_j), \tag{4.115}$$

where b is a real number such that $b > 1$. Then we obtain the upper bound:

$$p_N^{(n)}(\beta; a, \kappa) \leqslant \inf_{0<m_1\le\dots\le m_n<1} \sum_{i=1}^{n} \left[\frac{\kappa_i}{m_i}\ln(2) + \frac{\beta^2}{4} m_i a_i\right]. \tag{4.116}$$

Proof This will be along the lines of the proof of Proposition 4.14. Lemma 4.12 is applicable because for each $i = 1, \dots, n$ one has

$$b \prod_{j=1}^{i} \delta(\pi_j(\sigma), \pi_j(\sigma'))\delta(\alpha_j, \alpha'_j) \leqslant \prod_{j=1}^{i} \delta(\pi_j(\sigma), \pi_j(\sigma')) + (b-1) \prod_{j=1}^{i} \delta(\alpha_j, \alpha'_j). \tag{4.117}$$

Using Eq. (4.115), we have

$$\begin{aligned}
&\frac{1}{N}\mathbb{E}\left[\ln \sum_{\sigma,\,\alpha} w(\alpha)\exp[-\beta V(\sigma, \alpha)]\right] \\
&= \frac{1}{N}\mathbb{E}\left(\ln \sum_{\sigma,\,\alpha} \exp\left[\frac{y(\alpha_1)}{m_1}\right]\exp\left[\frac{y(\alpha_1, \alpha_2)}{m_2}\right]\cdots\exp\left[\frac{y(\alpha_1, \dots, \alpha_n)}{m_n}\right]\right. \\
&\quad \left.\times \exp\left[-\beta\sqrt{\frac{bN}{2}}\sum_{i=1}^{n}\sqrt{a_i}\, X(\pi_1(\sigma), \pi_2(\sigma), \dots, \pi_i(\sigma), \alpha_1, \alpha_2, \dots, \alpha_i)\right]\right).
\end{aligned} \tag{4.118}$$

Since the sum over configurations $\sigma \in \Sigma_N, \alpha \in \mathcal{A}^n$ can be decomposed into n sums over each subset $\pi_i(\sigma) \in \Sigma_{K_i}, \alpha_i \in \mathcal{A}$ for $i = 1, \dots, n$, the invariance property (4.64) can now be applied telescopically, starting at the nth level and tracing back up to the first level. After this simplification we obtain

$$\begin{aligned}
&\frac{1}{N}\mathbb{E}\left[\ln\sum_{\sigma,\alpha} w(\alpha)\exp[-\beta V(\sigma,\alpha)]\right] \\
&= \frac{1}{N}\mathbb{E}\left[\ln\sum_{\alpha} w(\alpha)\right] + \sum_{i=1}^{n}\frac{1}{m_i N}\ln\mathbb{E}\left[\mathcal{Z}_{K_i}^{m_i}(\beta\sqrt{ba_i})\right] \\
&\leq \frac{1}{N}\mathbb{E}\left[\ln\sum_{\alpha} w(\alpha)\right] + \sum_{i=1}^{n}\frac{1}{m_i N}\ln\mathbb{E}\left[\mathcal{Z}_{K_i}(m_i\beta\sqrt{ba_i})\right] \\
&= \frac{1}{N}\mathbb{E}\left[\ln\sum_{\alpha} w(\alpha)\right] + \sum_{i=1}^{n}\left[\frac{\kappa_i}{m_i}\ln(2)\right] + \frac{\beta^2}{4}b\sum_{i=1}^{n} a_i m_i
\end{aligned} \tag{4.119}$$

where in the third line we have again made use of Eq. (4.78).

Analogously, using Eq. (4.114) and the invariance property (4.64) we have

$$\begin{aligned}
&\frac{1}{N}\mathbb{E}\left[\ln\sum_{\alpha} w(\alpha)\exp[-\beta K(\alpha)]\right] \\
&\quad = \frac{1}{N}\mathbb{E}\left[\ln\sum_{\alpha} w(\alpha)\right] + \beta^2\frac{(b-1)}{4}\sum_{i=1}^{n} a_i m_i .
\end{aligned} \tag{4.120}$$

Putting together Eqs. (4.119) and (4.120), and optimizing over the choice of m_i, we arrive at the upper bound stated in the proposition. □

Let us proceed with the optimization in Eq. (4.116). In the following we make the assumption

$$\frac{\kappa_1}{a_1} < \frac{\kappa_2}{a_2} < \cdots < \frac{\kappa_n}{a_n} \tag{4.121}$$

in order to have a totally non-degenerate sequence of transition temperatures. To express the inequality of (4.116) in a more transparent form, it is convenient to introduce a succession of critical temperatures: for $i = 1, \ldots, n$ let $\beta_i^* = \beta_c\sqrt{\frac{\kappa_i}{a_i}}$ (where $\beta_c = 2\sqrt{\ln(2)}$ as in the REM). Under the condition (4.121), this implies that $\beta_1^* < \cdots < \beta_n^*$. Because of the constraint $0 < m_1 \leq \cdots \leq m_n < 1$, the optimal m_i is

$$m_i = \min\left\{1, \frac{\beta_i^*}{\beta}\right\} \tag{4.122}$$

for $i = 1, \dots, n$, the value 1 being attained by taking $m_i \uparrow 1$ in the infimum of Eq. (4.116). Therefore, one has

$$p_N^{(n)}(\beta; a, \kappa) \leq \bar{p}^{(n)}(\beta; a, \kappa), \tag{4.123}$$

where

$$\bar{p}^{(n)}(\beta; a, \kappa) = \begin{cases} \sum_{k=1}^{n} \frac{1}{4} a_k(\beta^2 + (\beta_k^*)^2) & \text{for } \beta < \beta_1^* \\ \sum_{k=1}^{i} \frac{1}{2} a_k \beta \beta_k^* + \sum_{k=i+1}^{n} \frac{1}{4} a_k(\beta^2 + (\beta_k^*)^2) & \text{for } \beta_i^* \leq \beta \leq \beta_{i+1}^* \\ \sum_{k=1}^{n} \frac{1}{2} a_k \beta \beta_k^* & \text{for } \beta \geq \beta_n^*. \end{cases} \tag{4.124}$$

The upper bound so obtained is a function made of parts of quadratic functions on each interval $[\beta_i, \beta_{i+1}]$. It is a continous function (together with its derivative) whose curvature starts from $1/2$ for $\beta < \beta_1$, is $\frac{1}{2} \sum_{k=i+1}^{n} a_k$ in the interval $[\beta_i, \beta_{i+1}]$, and is zero for $\beta > \beta_n$.

4.10 Lower bound for the GREM pressure

Let us denote the REM pressure (see Eq. (4.53)) as $p_N^{(1)}(\beta)$, and its thermodynamic limit (Eq. (4.54)) as $p^{(1)}(\beta)$. This is not really an abuse of notation because if $n = 1$ then the GREM is the REM, and $a_1 = \kappa_1 = 1$. In the same way, we write $\bar{p}^{(1)}(\beta) = \bar{p}(\beta)$, where $\bar{p}(\beta)$ is defined in Eq. (4.55).

Then the lower bound is the following.

Proposition 4.26 *For all $\beta \geq 0$,*

$$p^{(n)}(\beta; a, \kappa) \geq \bar{p}^{(n)}(\beta; a, \kappa). \tag{4.125}$$

Proof The proof will follow if we show that

$$p_N^{(n)}(\beta; a, \kappa) \geq \sum_{i=1}^{n} \kappa_i p_{K_i}^{(1)}\left(\sqrt{a_i/\kappa_i}\, \beta\right). \tag{4.126}$$

Indeed, taking the thermodynamic limit $N \to \infty$ on both sides and using

$$\bar{p}^{(n)}(\beta; a, \kappa) = \sum_{i=1}^{n} \kappa_i \bar{p}^{(1)}\left(\sqrt{a_i/\kappa_i}\, \beta\right) \tag{4.127}$$

we obtain the lemma statement. To prove (4.126) we introduce the interpolating pressure

$$\frac{1}{N} \mathbb{E} \ln \sum_{\sigma \in \Sigma_N} e^{-\beta \tilde{H}(\sigma, t)} \tag{4.128}$$

with

$$\tilde{H}(\sigma, t) = \sqrt{t}\left[\sqrt{\frac{N}{2}}\sum_{i=1}^{n}\sqrt{a_i}\,X(\pi_1(\sigma),\ldots,\pi_i(\sigma))\right] + \sqrt{1-t}\left[\sqrt{\frac{N}{2}}\sum_{i=1}^{n}\sqrt{a_i}\,Y(\pi_i(\sigma))\right] \tag{4.129}$$

where the Xs and Ys are families of i.i.d. standard Gaussian random variables, independent from each other. A straightforward differentiation of Eq. (4.128), combined with integration by parts, yields Eq. (4.126). □

4.11 The Aizenman–Sims–Starr extended variational principle

By exploiting the explicit expression of the Parisi functional for piecewise constant order parameters and the interpolation introduced in Guerra and Toninelli (2002), it was shown in Guerra (2003) that the Parisi free energy gives a lower bound for the free energy of the SK model. Instead of following that brilliant direct approach, we shall follow a more systematic path to get the same result. Namely, we will first describe an *extended variational principle* introduced in Aizenman *et al.* (2003), which provides a general method to obtain lower bounds for the free energy of mean-field models, as has been shown for the REMs and GREMs so far in this chapter. In this section we illustrate the method for the SK model by introducing general objects called *random overlap structures*. The Guerra (2003) bound will be obtained in Section 4.12 by the special choice of a hierarchical random overlap structure, together with the use of the Derrida–Ruelle random probability cascades.

The extended variational principle is inspired by the *cavity method* in physics. The *cavity field* was introduced in Onsager (1936) to describe the field that is experienced by the spin σ_i at site i. This field is different from the instantaneous field on the site i, given by $\sum_{j=1}^{N} J_{i,j}\sigma_j$. In particular, while the instantaneous fields are correlated for different sites, the cavity fields are independent for different sites (in fact, they are constructed by analyzing the effect of removing the spin at site i, thus the name *cavity field*). The idea of the method is therefore to compare the free energies for a system of N spins and a system of $N+1$ spins. The existence of the thermodynamic limit (via subadditivity) implies that the differences between free energies can be controlled. The application of the method to spin glasses is discussed thoroughly in Mézard *et al.* (1987).

From a mathematical point of view, it is convenient to consider the incremental free energy when one passes from a system of size N to a system of size $N+M$. After the limit $N \to \infty$ is taken, this leads to the construction of an auxiliary system which is placed in relation to the original system of size M using the comparison inequality proved in Theorem 3.46. After a subsequent limit $M \to \infty$ is taken, the auxiliary system will act as a forcing one which will also impose the value of its overlap distribution on the original system (in a similar way, a thermal reservoir will impose its temperature on a bulk system).

The general construction of auxiliary systems requires the introduction of the concept of random overlap structures and the cavity functional (we will follow Aizenman *et al.* (2003)). Let us denote by $\alpha \in \mathcal{A}$, an index belonging to the denumerable set $\mathcal{A}$ which collects all the possible configurations of the auxiliary system. One assumes the existence of an overlap $p_{\alpha,\alpha'}$ between two configurations α and α' of the auxiliary system, which is a positive definite matrix. Moreover one assumes that the auxiliary system carries normalizable random weights ξ_α (in the general case the overlap can also be random). Therefore one has the following definition.

Definition 4.27 (Random overlap structures) The couple $r = (p, \xi)$, given by the covariance matrix $p_{\alpha,\alpha'}$ and the stochastic process ξ_α with law μ defines a *random overlap structure* (ROSt) if

1. $\xi_\alpha \geqslant 0$ and $\sum_{\alpha \in \mathcal{A}} \xi_\alpha < \infty$ μ-almost surely;
2. $p_{\alpha,\alpha'}$ is a positive definite matrix;
3. $p_{\alpha,\alpha} = 1$, which implies by Schwartz inequality $p_{\alpha,\alpha'} \leqslant 1$.

We will denote by $\mathcal{R}$ the set of all ROSt with a given covariance matrix $p_{\alpha,\alpha'}$ and law μ for the random weights ξ_α.

To define the cavity functional which will appear in the pressure upper bound, it is convenient to associate a given ROSt to two independent families of centered Gaussian random variables $\{K_\alpha\}$ and $\{\eta_{j,\alpha}\}_{j=1,2,\dots}$ with covariances

$$\mathbb{E}(\eta_{j,\alpha}\eta_{j',\alpha'}) = \delta_{j,j'}\, p_{\alpha,\alpha'} \tag{4.130}$$

$$\mathbb{E}(K_\alpha K_{\alpha'}) = p^2_{\alpha,\alpha'}. \tag{4.131}$$

The cavity functional can now be defined in terms of a given ROSt.

Definition 4.28 (Cavity functional) For a spin system of size M, with spin configuration $\sigma \in \{-1,+1\}^M$, and a general ROSt $r \in \mathcal{R}$, the cavity functional is defined as

$$G_{M,r}(\beta, h) = \frac{1}{M}\mathbb{E}\left(\log\left(\frac{\sum_{\sigma,\alpha} \xi_\alpha\, e^{\beta \sum_{j=1}^{M}(\eta_{j,\alpha}+h)\sigma_j}}{\sum_\alpha \xi_\alpha e^{\beta\sqrt{\frac{M}{2}}K_\alpha}}\right)\right). \tag{4.132}$$

In the above definition, $\mathbb{E}$ denotes expectation with respect to all the randomness involved, i.e. the Gaussian families K and η, and the random weight ξ.

In Aizenman *et al.* (2003) it is proved that in the thermodynamic limit, the pressure of the SK model is obtained by minimization of the cavity functional, i.e. the free energy is obtained by maximization of the cavity functional (akin to the maximization that is invoked in the Parisi solution).

Theorem 4.29 (Extended variational principle) *The infinite-volume pressure of the SK model satisfies*

$$p^{(SK)}(\beta, h) = \lim_{M\to\infty} \inf_{r\in\mathcal{R}} G_{M,r}(\beta, h). \tag{4.133}$$

Proof The proof of the theorem is divided into two steps. First, the upper bound it is proved, valid for all M,

$$p_M^{(SK)}(\beta, h) \leqslant \inf_{r\in\mathcal{R}} G_{M,r}(\beta, h). \tag{4.134}$$

This is shown in Proposition 4.30 by making use of the comparison inequality in Theorem 3.46. It follows that the same bound is also clearly true in the thermodynamic limit, i.e.

$$p^{(SK)}(\beta, h) \leqslant \lim_{M\to\infty} \inf_{r\in\mathcal{R}} G_{M,r}(\beta, h). \tag{4.135}$$

To complete the proof, the opposite inequality is also needed. This is proved in Proposition 4.31, exploiting subadditivity of the pressure proved by Guerra and Toninelli and the Polya–Szego Theorem. □

Proposition 4.30 (Upper bound extended variational principle) *The finite-volume pressure of the SK model satisfies, for all $M \in \mathbb{N}$,*

$$p_M^{(SK)}(\beta, h) \leqslant \inf_{r\in\mathcal{R}} G_{M,r}(\beta, h). \tag{4.136}$$

Proof It is enough to show that the finite-volume pressure is upper-bounded by the cavity functional

$$p_M^{(SK)}(\beta, h) \leqslant G_{M,r}(\beta, h). \tag{4.137}$$

Equation (4.136) then follows by optimizing (i.e. taking the infimum) over all possible random overlap structures.

Equation (4.137) can be rewritten as

$$\mathbb{E}\left(\log\left(\sum_{\sigma,\alpha} \xi_\alpha\, e^{-\beta(H_\sigma^{(SK)} - h\sum_{i=1}^M \sigma_i)}\, e^{\beta\sqrt{\frac{M}{2}}K_\alpha}\right)\right)$$

$$\leqslant \mathbb{E}\left(\log\left(\sum_{\sigma,\alpha} \xi_\alpha\, e^{\beta\sum_{i=1}^M (\eta_{i,\alpha}+h)\sigma_i}\right)\right) \tag{4.138}$$

The above inequality is a consequence of the comparison inequality in Theorem 3.46, applied to the two processes

$$X_{\sigma,\alpha} = -\beta H_{\sigma}^{(SK)} + \beta\sqrt{\frac{M}{2}}K_{\alpha} \tag{4.139}$$

$$Y_{\sigma,\alpha} = \beta\sum_{i=1}^{M}\eta_{i,\alpha}\sigma_i. \tag{4.140}$$

Indeed, we have

$$\mathbb{E}(X_{\sigma,\alpha}X_{\sigma',\alpha'}) = \beta^2\frac{M}{2}q_{\sigma,\sigma'}^2 + \beta^2\frac{M}{2}p_{\alpha,\alpha'}^2 \tag{4.141}$$

$$\mathbb{E}(Y_{\sigma,\alpha}Y_{\sigma',\alpha'}) = \beta^2 M q_{\sigma,\sigma'}p_{\alpha,\alpha'}, \tag{4.142}$$

and since

$$q_{\sigma,\sigma'}^2 + p_{\alpha,\alpha'}^2 \geqslant 2q_{\sigma,\sigma'}p_{\alpha,\alpha'}, \tag{4.143}$$

we have

$$\mathbb{E}(X_{\sigma,\alpha}X_{\sigma',\alpha'}) \geqslant \mathbb{E}(Y_{\sigma,\alpha}Y_{\sigma',\alpha'}). \tag{4.144}$$

□

Proposition 4.31 (Lower bound extended variational principle) *The infinite-volume pressure of the SK model satisfies*

$$p^{(SK)}(\beta,h) \geqslant \lim_{M\to\infty}\inf_{r\in\mathcal{R}} G_{M,r}(\beta,h). \tag{4.145}$$

Proof To prove the proposition we use the following lemma by Polya–Szego (Lemma B.1 of Aizenman *et al.* (2007) also provides a proof).

Lemma 4.32 (Superadditive sequences limit) *If a sequence* $\{P_N\}_{N\in\mathbb{N}}$ *is superadditive, i.e. for each* $N, M \in \mathbb{N}$

$$P_{N+M} \geqslant P_N + P_M \tag{4.146}$$

then the following limits exist and satisfy

$$\lim_{N\to\infty}\frac{P_N}{N} = \lim_{M\to\infty}\liminf_{N\to\infty}\frac{[P_{N+M}-P_N]}{M}. \tag{4.147}$$

The above lemma can be applied in our setting because the sequence of the SK pressure $\{P_N\}_{N\in\mathbb{N}}$ has been shown to be superadditive in Guerra and Toninelli (2002) (see also Examples 3.50–3.53). Using Lemma 4.32, inequality (4.145) is equivalent to

$$\lim_{M\to\infty}\liminf_{N\to\infty}\frac{[P_{N+M}-P_N]}{M}\geqslant\lim_{M\to\infty}\inf_{r\in\mathcal{R}}G_{M,r}(\beta,h). \tag{4.148}$$

This inequality would follow if we could prove that, for every M:

$$\liminf_{N\to\infty}\frac{[P_{N+M}-P_N]}{M}\geqslant\inf_{r\in\mathcal{R}}G_{M,r}(\beta,h). \tag{4.149}$$

To show this, it is enough to exhibit one particular random overlap structure $r^*\in\mathcal{R}$ such that

$$\liminf_{N\to\infty}\frac{1}{M}\mathbb{E}\left(\log\frac{\mathcal{Z}_{N+M}}{\mathcal{Z}_N}\right)=G_{M,r^*}(\beta,h). \tag{4.150}$$

The construction of this particular ROSt is based on the fact that – up to corrections that vanish in the limit – when a system of M spins $(\sigma_1,\ldots,\sigma_M)$ is added to a much larger system of N spins $\alpha=(\alpha_1,\ldots\alpha_N)$, in particular from now on $\alpha_i=\{-1,+1\}$, the change in pressure is exactly of the form appearing in the cavity functional (4.132). In the limit of large volumes, the larger block acts as a ROSt on the smaller volume. Somehow the larger block imposes its overlap distribution on the smaller volume, similar to the way in which an infinite thermal reservoir fixes the temperature of a smaller system attached to it.

To describe the peculiar ROSt r^* we compare the pressures of a system of size $N+M$ with a system of size N

$$\mathbb{E}\left(\log\frac{\mathcal{Z}_{M+N}}{\mathcal{Z}_N}\right)=\mathbb{E}\left(\log\frac{\sum_{\alpha,\sigma}e^{-\beta H_{N+M}(\alpha,\sigma)}}{\sum_{\alpha}e^{-\beta H_N(\alpha)}}\right). \tag{4.151}$$

We then split the total Hamiltonian of size $N+M$ into three parts:

1. the interaction within the larger block of size N;
2. the interaction between spins in the larger block and spins in the smaller block of size M;
3. the interaction within the smaller block of size M:

$$H_{N+M}(\alpha,\sigma)=\frac{1}{\sqrt{2(N+M)}}\sum_{i=1}^{N}\sum_{j=1}^{N}J_{i,j}\alpha_i\alpha_j+h\sum_{i=1}^{N}\alpha_i \tag{4.152}$$

$$+\frac{1}{\sqrt{N+M}}\sum_{i=1}^{N}\sum_{j=1}^{M}J_{i,j}\alpha_i\sigma_j+h\sum_{j=1}^{M}\sigma_j \tag{4.153}$$

$$+\frac{1}{\sqrt{2(N+M)}}\sum_{i=1}^{M}\sum_{j=1}^{M}J_{i,j}\sigma_i\sigma_j. \tag{4.154}$$

The first term in the above expression

$$\tilde{H}_N(\alpha) = \frac{1}{\sqrt{2(N+M)}} \sum_{i=1}^{N} \sum_{j=1}^{N} J_{i,j} \alpha_i \alpha_j \tag{4.155}$$

looks like the Hamiltonian of the SK model of size N (see Definition 1.3). The difference arises from the normalization. However, by the law of addition of independent Gaussian random variables, we have the following equality in distribution

$$H_N(\alpha) \stackrel{\mathcal{D}}{=} \tilde{H}_N(\alpha) + \sqrt{\frac{M}{2(N+M)N}} \sum_{i=1}^{N} \sum_{j=1}^{N} J_{i,j} \alpha_i \alpha_j. \tag{4.156}$$

Calling $U(\sigma, \alpha)$ the term in Eq. (4.154) we can rewrite the partition function ratio in (4.151) as

$$\begin{aligned} &\mathbb{E}\left(\log \frac{\mathcal{Z}_{M+N}}{\mathcal{Z}_N}\right) \\ &\quad = \mathbb{E}\left(\log \frac{\sum_{\alpha,\sigma} e^{\beta\left(\tilde{H}(\alpha)+h\sum_{i=1}^{N}\alpha_i\right)} e^{\beta\sum_{j=1}^{M}\sigma_j\left(\frac{1}{\sqrt{N+M}}\sum_{i=1}^{N} J_{i,j}\alpha_i + h\right)} e^{\beta U(\sigma,\alpha)}}{\sum_{\alpha} e^{\beta\left(\tilde{H}(\alpha)+h\sum_{i=1}^{N}\alpha_i\right)} e^{\beta\sqrt{\frac{M}{2}}\sqrt{\frac{1}{(N+M)N}}\sum_{i,j=1}^{N} J_{i,j}\alpha_i\alpha_j}}\right) \end{aligned} \tag{4.157}$$

For the construction of the particular ROSt r^*, this suggests to take the random weights

$$\xi^*_\alpha = e^{\beta\left(\tilde{H}(\alpha)+h\sum_{i=1}^{N}\alpha_i\right)} \tag{4.158}$$

and the overlap kernel

$$p^*_{\alpha,\alpha'} = \frac{1}{N} \sum_{i=1}^{N} \alpha_i \alpha'_i. \tag{4.159}$$

The fields $\{\eta^*_{j,\alpha}\}$ and $\{K^*_\alpha\}$ are identified with

$$\eta^*_{j,\alpha} = \frac{1}{\sqrt{N+M}} \sum_{i=1}^{N} J_{i,j} \alpha_i \tag{4.160}$$

$$K^*_\alpha = \sqrt{\frac{1}{(N+M)N}} \sum_{i,j=1}^{N} J_{i,j} \alpha_i \alpha_j \tag{4.161}$$

and are such that

$$\mathbb{E}(\eta^*_{j,\alpha} \eta^*_{j',\alpha'}) = \frac{N}{N+M} \delta_{j,j'}\, p^*_{\alpha,\alpha'} \tag{4.162}$$

$$\mathbb{E}(K^*_\alpha K^*_{\alpha'}) = \frac{N}{N+M} (p^*_{\alpha,\alpha'})^2. \tag{4.163}$$

Therefore in the limit $N \to \infty$, the fields' covariances satisfy (4.131) and (4.130). Moreover, the field U is negligible in the limit $N \to \infty$ because

$$\mathbb{E}(U^2(\alpha, \sigma)) = \frac{M^2}{2N}. \tag{4.164}$$

Therefore Eq. (4.150) is established and this completes the proof of Proposition 4.31. □

4.12 Guerra upper bound

We are now in the position to present the result in Guerra (2003), whose origin can be found in Guerra (1995) – namely, the Parisi pressure defined in Proposition 1.16 provides an upper bound for the pressure of the SK model.

Theorem 4.33 (Guerra upper bound) *The finite-volume pressure per particle of the SK model (cf. Definition 1.3) is upper bounded uniformly in the system size by the Parisi functional (1.86) with an arbitrary functional order parameter $x(q) \in \mathcal{M}$, i.e.*

$$p_N^{(\mathrm{SK})}(\beta, h) \leqslant \mathcal{P}(x(q), \beta, h). \tag{4.165}$$

Therefore, in the limit $N \to \infty$, one has

$$p^{(\mathrm{SK})}(\beta, h) \leqslant p^{(\mathrm{Parisi})}(\beta, h) \tag{4.166}$$

where $p^{(\mathrm{Parisi})}(\beta, h)$ is given in (1.85).

Proof By the extended variational principle – Theorem 4.29 – it suffices to exhibit a particular ROSt $\bar{r} \in \mathcal{R}$ such that the cavity functional coincides with the Parisi functional, i.e.

$$G_{M,\bar{r}}(\beta, h) = \mathcal{P}(x(q), \beta, h). \tag{4.167}$$

Indeed, if this is the case, since Eq. (4.137) in Proposition 4.30 is valid for any $r \in \mathcal{R}$, it implies in particular that

$$p_M^{(\mathrm{SK})}(\beta, h) \leqslant G_{M,\bar{r}}(\beta, h). \tag{4.168}$$

This proves (4.165). It is then convenient to optimize (i.e. take the infimum) over the functional order parameter $x(q)$ and take the limit $M \to \infty$ to establish (4.166).

We now describe the construction of the particular ROSt $\bar{r}$ and then verify that the cavity functional on this ROSt does coincide with the Parisi functional. An important class of ROSt is obtained when the law of the random weights ξ_α is chosen to be the random probability cascades introduced by Ruelle in his description of the Derrida GREM (see Ruelle (1987); Derrida (1985); Derrida and Gardner (1986)). This structure is obtained as a hierarchical construction, which combines the exponential point process introduced for the description of the energy level statistics of the REM in the low temperature phase, with the piecewise order parameter introduced by Parisi.

To describe the construction, a multi-index $\alpha = (\alpha_1, \alpha_2, \ldots, \alpha_k)$, is needed with $k \in \mathbb{N}$ and $\alpha_i \in \mathcal{A}$ for $1 \leqslant i \leqslant k$. The set $\mathcal{A}$ is a countably infinite set and we may, for instance, take the set $\mathbb{N}$ of natural integers. For a given piecewise constant order parameter

$$x(q) = \begin{cases} m_1 & \text{if } 0 = q_0 \leqslant q < q_1 \\ m_2 & \text{if } q_1 \leqslant q < q_2 \\ \vdots & \\ m_k & \text{if } q_{k-1} \leqslant q \leqslant q_k = 1 \end{cases} \tag{4.169}$$

specified by the monotone sequences

$$0 = m_1 \leqslant m_2 \leqslant \ldots \leqslant m_{k-1} \leqslant m_k = 1, \tag{4.170}$$

the random probability cascades process is constructed in the following manner:

1. Start with $\{X_{\alpha_1}\}_{\alpha_1 \in \mathcal{A}}$, a Poisson point process with intensity measure $e^{-x}dx$, and define $Y_{\alpha_1} = e^{\frac{X_{\alpha_1}}{m_1}}$. Therefore $\{Y_{\alpha_1}\}_{\alpha_1 \in \mathcal{A}}$ constitutes a Poisson point process of intensity $m_1 y^{-m_1-1}dy$. The points of the process, labeled by α_1, are not ordered.
2. For each $\alpha_1 \in \mathcal{A}$, we generate a Poisson point process $\{X_{\alpha_1,\alpha_2}\}_{\alpha_2 \in \mathcal{A}}$ having intensity measure $e^{-x}dx$, and then define $Y_{\alpha_1,\alpha_2} = e^{\frac{X_{\alpha_1,\alpha_2}}{m_2}}$. Therefore $\{Y_{\alpha_1,\alpha_2}\}_{\alpha_2 \in \mathcal{A}}$ constitutes a Poisson point process of intensity $m_2 y^{-m_2-1}dy$. The processes corresponding to different values of α_1 are chosen independently.
3. The construction is iterated k times. At the i th step, with $i \leqslant k$, for each $(\alpha_1, \ldots, \alpha_{i-1})$, the process $\{X_{\alpha_1,\ldots,\alpha_{i-1},\alpha_i}\}_{\alpha_i \in \mathcal{A}}$ constitutes an independent copy of the exponential Poisson point process and $\{Y_{\alpha_1,\ldots,\alpha_{i-1},\alpha_i}\}_{\alpha_i \in \mathcal{A}}$ is an independent Poisson point process of intensity $m_i y^{-m_i-1}dy$.

Having defined the random probability cascades process, the set of weights ξ_α of the hierarchical ROSt $\bar{r}$ are given by the product of the components of the process, i.e.

$$\xi_\alpha = Y_{\alpha_1} Y_{\alpha_1,\alpha_2} \cdots Y_{\alpha_1,\ldots,\alpha_k} \tag{4.171}$$

The other ingredient needed to fully specify the ROSt $\bar{r}$, is the overlap kernel $\bar{p}_{\alpha,\alpha'}$. This is chosen to have an ultrametric structure, i.e.

$$\bar{p}_{\alpha,\alpha'} = \begin{cases} q_0 = 0 & \text{if } \alpha_1 \neq \alpha_1' \\ q_1 & \text{if } \alpha_1 = \alpha_1', \alpha_2 \neq \alpha_2' \\ q_2 & \text{if } \alpha_1 = \alpha_1', \alpha_2 = \alpha_2', \alpha_3 \neq \alpha_3' \\ \vdots & \\ q_k = 1 & \text{if } \alpha_1 = \alpha_1', \alpha_2 = \alpha_2', \ldots, \alpha_k = \alpha_k'. \end{cases} \tag{4.172}$$

We recall that the two independent Gaussian fields $\bar{\eta}_{j,\alpha}$ and $\bar{K}_\alpha$ entering the expression of the cavity functional $G_{M,r}$ are defined by Eqs. (4.130) and (4.131); that is

$$\mathbb{E}(\bar{\eta}_{j,\alpha}\bar{\eta}_{j',\alpha'}) = \delta_{j,j'}\, \bar{p}_{\alpha,\alpha'} \tag{4.173}$$

$$\mathbb{E}(\bar{K}_\alpha \bar{K}_{\alpha'}) = \bar{p}^2_{\alpha,\alpha'}. \tag{4.174}$$

An explicit construction of those fields is given by

$$\bar{\eta}_{j,\alpha} = \sqrt{q_1 - q_0}\, J_{j,\alpha_1} + \sqrt{q_2 - q_1}\, J_{j,\alpha_1,\alpha_2} + \cdots + \sqrt{q_k - q_{k-1}}\, J_{j,\alpha_1,\ldots,\alpha_k} \tag{4.175}$$

$$\bar{K}_\alpha = \sqrt{q_1^2 - q_0^2}\, \tilde{J}_{\alpha_1} + \sqrt{q_2^2 - q_1^2}\, \tilde{J}_{\alpha_1,\alpha_2} + \cdots + \sqrt{q_k^2 - q_{k-1}^2}\, \tilde{J}_{\alpha_1,\ldots,\alpha_k} \tag{4.176}$$

where $J_{j,\alpha_1}, J_{j,\alpha_1,\alpha_2}, \ldots, J_{j,\alpha_1,\alpha_2,\ldots,\alpha_k}$ and $\tilde{J}_{\alpha_1}, \tilde{J}_{\alpha_1,\alpha_2}, \ldots, \tilde{J}_{\alpha_1,\alpha_2,\ldots,\alpha_k}$ are all independent standard Gaussian random variables. One can immediately verify that Eqs. (4.173) and (4.174) are indeed satisfied.

To complete the proof of the theorem, we need to show that indeed

$$G_{N,\bar{r}}(\beta, h) = \mathcal{P}(x(q), \beta, h). \tag{4.177}$$

To show this, we prove below that

$$\frac{1}{M}\mathbb{E}\left(\log\left(\sum_{\sigma,\alpha} \xi_\alpha\, e^{\beta \sum_{j=1}^{M} (\bar{\eta}_{j,\alpha}+h)\sigma_j}\right)\right) = \log 2 + f_{x(q)}(0,h) + \mathbb{E}\log\sum_\alpha \xi_\alpha, \tag{4.178}$$

$$\frac{1}{M}\mathbb{E}\left(\log\left(\sum_\alpha \xi_\alpha e^{\beta\sqrt{\frac{M}{2}}\bar{K}_\alpha}\right)\right) = \frac{\beta^2}{2}\int_0^1 q\, x(q) dq + \mathbb{E}\log\sum_\alpha \xi_\alpha. \tag{4.179}$$

When we take the difference (side by side) of the above equations, and using Definition 4.28 of the cavity functional $G_{M,r}$ and Definition 1.86 of the Parisi functional, we see that $G_{M,\bar{r}}(\beta, h) = \mathcal{P}(x(q), \beta, h)$.

To prove the two equations above, we use the explicit representation of the Parisi functional in the case of piecewise constant order parameter $x(q)$, and the crucial invariance property of the exponential Poisson point process which was proved in Proposition 4.3.

Let us start from Eq. (4.178). Summing over the spins σ, we cast the lefthand side in the form

$$\ln(2) + \frac{1}{M}\mathbb{E}\left(\log\left(\sum_{\alpha}\bar{\xi}_{\alpha}\prod_{j=1}^{M}e^{\ln(\cosh(\beta(\bar{\eta}_{j,\alpha}+h)))}\right)\right). \tag{4.180}$$

The hierarchical random probability cascade inherits the remarkable invariance property of its building block, the exponential Poisson point process. It is convenient to define the partial quantities, for $l = 1, \ldots, k$

$$\bar{\xi}^{(l)}_{\alpha_1,\ldots,\alpha_l} = \prod_{i=1}^{l} Y_{\alpha_1,\ldots,\alpha_i} \tag{4.181}$$

and

$$\bar{\eta}^{(l)}_{j,\alpha_1,\ldots,\alpha_l} = \sum_{i=0}^{l}\sqrt{q_{i+1}-q_i}\; J_{j,\alpha_1,\ldots,\alpha_i}. \tag{4.182}$$

Note that for $l = k$, we have $\bar{\xi}^{(k)}_{\alpha_1,\ldots,\alpha_k} = \bar{\xi}_{\alpha}$ and $\bar{\eta}^{(k)}_{j,\alpha_1,\ldots,\alpha_k} = \bar{\eta}_{j,\alpha}$. Conditioning on the collection of variables $\bar{\xi}^{(k-1)}_{\alpha_1,\ldots,\alpha_{k-1}}$ and $\bar{\eta}^{(k-1)}_{j,\alpha_1,\ldots,\alpha_{k-1}}$ and for each $\alpha_1, \ldots, \alpha_{k-1}$, we can use Eq. (4.8) to obtain

$$\begin{aligned}
&\frac{1}{M}\mathbb{E}\left(\log\left(\sum_{\alpha}\bar{\xi}_{\alpha}\prod_{j=1}^{M}e^{\ln(\cosh(\beta(\bar{\eta}_{j,\alpha}+h)))}\right)\right)\\
&= \frac{1}{M}\mathbb{E}\left(\log\left(\sum_{\alpha_1,\ldots,\alpha_{k-1}}\bar{\xi}^{(k-1)}_{\alpha_1,\ldots,\alpha_{k-1}}\sum_{\alpha_k}Y_{\alpha_1,\ldots,\alpha_k}\right.\right.\\
&\qquad\left.\left.\times\prod_{j=1}^{M}e^{\ln(\cosh(\beta(\bar{\eta}^{(k-1)}_{j,\alpha_1,\ldots,\alpha_{k-1}}+\sqrt{q_k-q_{k-1}}J_{j,\alpha_1,\ldots,\alpha_k}+h)))}\right)\right)\\
&= \frac{1}{M}\mathbb{E}\left(\log\left(\sum_{\alpha_1,\ldots,\alpha_{k-1}}\bar{\xi}^{(k-1)}_{\alpha_1,\ldots,\alpha_{k-1}}\sum_{\alpha_k}Y_{\alpha_1,\ldots,\alpha_k}\right.\right.\\
&\qquad\left.\left.\times\prod_{j=1}^{M}\left(\mathbb{E}_{j,k}\left(e^{m_k\ln(\cosh(\beta(\bar{\eta}^{(k-1)}_{j,\alpha_1,\ldots,\alpha_{k-1}}+\sqrt{q_k-q_{k-1}}X_{j,k}+h)))}\right)\right)^{1/m_k}\right)\right)
\end{aligned}$$

where $\mathbb{E}_{j,k}$ denotes expectation with respect to the jth of the M i.i.d. standard Gaussian random variables $X_{j,k}$. By iterating this computation and using the invariance property of the exponential Poisson process at each level of the hierarchy, one arrives at (4.178).

In a similar way, one proves Eq. (4.179). Defining for $l = 1, \dots, k$

$$\bar{K}^{(l)}_{\alpha_1,\dots,\alpha_l} = \sum_{i=0}^{l} \sqrt{q_{i+1}^2 - q_i^2}\; \tilde{J}_{\alpha_1,\dots,\alpha_i}, \tag{4.183}$$

conditioning on the collection of variables $\bar{\xi}^{(k-1)}_{\alpha_1,\dots,\alpha_{k-1}}$ and $\bar{K}^{(k-1)}_{j,\alpha_1,\dots,\alpha_{k-1}}$, and for each $\alpha_1, \dots, \alpha_{k-1}$, the first step of the iteration reads:

$$\begin{aligned}
&\frac{1}{M}\mathbb{E}\left(\log\left(\sum_{\alpha} \bar{\xi}_\alpha e^{\beta\sqrt{\frac{M}{2}}\bar{K}_\alpha}\right)\right) \\
&= \frac{1}{M}\mathbb{E}\left(\log\left(\sum_{\alpha_1,\dots,\alpha_{k-1}} \bar{\xi}^{(k-1)}_{\alpha_1,\dots,\alpha_{k-1}} \bar{K}^{(k-1)}_{\alpha_1,\dots,\alpha_{k-1}} \sum_{\alpha_k} Y_{\alpha_1,\dots,\alpha_k} e^{\beta\sqrt{\frac{M}{2}}\sqrt{q_k^2-q_{k-1}^2}\tilde{J}_{\alpha_1,\dots,\alpha_k}}\right)\right) \\
&= \frac{1}{M}\mathbb{E}\left(\log\left(\sum_{\alpha_1,\dots,\alpha_{k-1}} \bar{\xi}^{(k-1)}_{\alpha_1,\dots,\alpha_{k-1}} \bar{K}^{(k-1)}_{\alpha_1,\dots,\alpha_{k-1}}\right.\right. \\
&\qquad \left.\left. \times \sum_{\alpha_k} Y_{\alpha_1,\dots,\alpha_k}\left(\mathbb{E}\left(e^{m_k\beta\sqrt{\frac{M}{2}}\sqrt{q_k^2-q_{k-1}^2}\tilde{J}_{\alpha_1,\dots,\alpha_k}}\right)\right)^{1/m_k}\right)\right) \\
&= \frac{1}{M}\mathbb{E}\left(\log\left(\sum_{\alpha_1,\dots,\alpha_{k-1}} \bar{\xi}^{(k-1)}_{\alpha_1,\dots,\alpha_{k-1}} \bar{K}^{(k-1)}_{\alpha_1,\dots,\alpha_{k-1}} \sum_{\alpha_k} Y_{\alpha_1,\dots,\alpha_k}\right)\right) + \frac{\beta^2}{4} m_k\left(q_k^2 - q_{k-1}^2\right).
\end{aligned}$$

□

The methods developed in Guerra's comparison strategy (done by interpolation and the successive general picture given by the extended variational principle of Aizenman–Sims–Starr) have found many fruitful applications to prove bounds for disordered models, including the Viana–Bray model (Franz *et al.* (2003)), the comparison between disordered Kac and SK models (Franz and Toninelli (2004)), random energy and generalized random energy models (Giardinà and Starr (2007)), and the antiferromagnet on the Erdős–Rényi random graph (Contucci *et al.* (2011a)).

4.13 Talagrand theorem and open problems

We have seen that the Guerra bound proves that the Parisi free energy expression for the SK model is a lower bound for the real value. The following theorem, which was proved after the Guerra work, establishes that the Parisi solution is indeed the exact value for the free energy density in the thermodynamic limit.

Theorem 4.34 (Talagrand (2006)) *The infinite-volume pressure per particle of the SK model is given by the Parisi pressure*

$$p^{(SK)}(\beta, h) = p^{(Parisi)}(\beta, h). \tag{4.184}$$

For the proof we refer to the original paper (Talagrand (2006)). The Guerra expression for the lower bound explicitly identifies a remainder which Talagrand proves vanishing in the thermodynamic limit. In the framework of the extended variational principle, this shows that when the reservoir is a hierarchical ROSt, the correction terms vanish in the thermodynamic limit – implying complete locking between the original and the auxiliary ultrametric systems.

The Guerra–Talagrand result has rigorously solved the longstanding conjecture of the Parisi solution as far as the free energy density is concerned. For the SK model, as for any other statistical mechanics model, the complete solution involves the whole characterization of the equilibrium state in the thermodynamic limit. The core problem for the structural property of the quenched measure for the SK model is the ultrametricity property, which is still an open problem (Bolthausen and Kistler (2009)). Although the ultrametric structure appears in the Guerra–Talagrand result, it is not proved that the equilibrium state has the ultrametric property as formulated in (1.48).

It is important to quote a few results that hint toward a possible solution of the problem. Following the competing particle approach with invariance under reshuffling (Arguin (2008); Arguin and Aizenman (2009)), and the spin model approach satisfying the Ghirlanda–Guerra identities (Panchenko (2010a); Talagrand (2010a)), these works show that ultrametricity holds under the hypothesis that the overlap is a discrete random variable. The result holds true independently of the specific model considered.

For the SK model in finite volume one has a discrete overlap distribution, but the Ghirlanda–Guerra identities are fulfilled only up to a finite-volume correction. On the other hand, in the thermodynamic limit when the correction to the identities vanishes, the overlap distribution becomes continuous. It is still an open problem to find a suitable scheme to apply the previous strategy to prove ultrametricity. See the recent preprint Panchenko (2011a) where the problem of ultrametricity is studied for a class of mean-field models.

5

Spin glass identities

Abstract

In this chapter we analyze some properties of the overlap probability distribution by looking at its factorization rules. It has been now understood that the disputed features of the spin glass phase are, apart from the triviality issue of the single overlap distribution, related to the structure of the joint overlap distribution. For example, a centered Gaussian family distribution is completely identified by its covariance, thanks to the Wick theorem factorization law. In the spin glass phase, the factorization structure first appeared within the Parisi replica symmetry breaking solution of the Sherrington–Kirkpatrick model. In this chapter we show that some of those properties can indeed be recovered by using a stability argument which amounts to proving how the equilibrium state is left unchanged by small perturbations. In turn, this is equivalent to controlling the size of the fluctuations for suitable thermodynamic quantities. After introducing the general method we review the stochastic stability, the control of thermal fluctuations, and the graph-theoretical description of the emerging identities. We then analyze the fluctuations due to the disorder and establish the self-averaging of the random internal energy. The extension to interactions with non-zero averages and the specific analysis on the Nishimori line follows. The chapter ends with new identities derived from the control of the fluctuations for free energy differences involving flips of the interactions.

5.1 The stability method and the structural identities

The solution of the Sherrington–Kirkpatrick model obtained via the replica approach showed that the order parameter of the theory, namely the overlap distribution, is sufficient to fully describe the quenched state. Using that approach, in the limit in which the number of replicas n is sent to zero, *combinatorial identities* among overlap functions were derived in Mézard *et al.* (1984) under the assumptions of replica equivalence and ultrametricity for the matrix Q introduced in Chapter 1 (see also Parisi (2004)).

In the genuine probabilistic approach, the former combinatorial identities can be translated into factorization properties of the joint distributions of overlaps among real replicas. In this chapter we illustrate a unified treatment to derive some of those structural identities, which only involves probabilistic methods on rigorous grounds. The primary tool in the analysis will be the stability of the quenched Boltzmann–Gibbs measure under a suitable class of perturbations.

The linear response stability under small perturbations is a basic method in statistical physics (see, for instance, Kubo *et al.* (1991); Landau and Lifshitz (1969)). Its meaning is conveyed by a simple idea: if a system is stable with respect to a suitable perturbation (small deformation) of its energy, then – in the thermodynamic limit – structural identities among physical quantities can be obtained by simply equating a derivative to zero. Those identities single out the order parameters of the theory, thus reducing its a priori degrees of freedom. The mean-field theory for the ferromagnetic phase in which the only order parameter is the magnetization, is a celebrated example of such an approach.

The first stability theory for the spin glass phase (called *stochastic stability*) was introduced in Aizenman and Contucci (1998) for the Sherrington–Kirkpatrick model, where it was shown that an infinite family of polynomial identities were implied for the overlap (see also Barra (2006), at equilibrium; Franz *et al.* (1998, 1999), out of equilibrium). A larger set of identities for the same model, known as the Ghirlanda–Guerra identities, was later obtained in Ghirlanda and Guerra (1998) through the bound of the energy fluctuations with respect to the quenched equilibrium state. A fully rigorous extension of the previous results to general Gaussian spin glass models, including short-range finite-dimensional models, was proved in Contucci and Giardinà (2005, 2007) and generalized in Contucci *et al.* (2011c).

Let us start by illustrating the idea of stability with classical deterministic spin systems. The Gibbs–Boltzmann state $\omega_{\beta,\Lambda}$ of a statistical mechanics system of $|\Lambda|$ interacting spins $\sigma = (\sigma_1, \dots, \sigma_{|\Lambda|})$, with Hamiltonian $H_\Lambda(\sigma)$ at inverse temperature β, admits the classical probabilistic interpretation as the deformation of the uniform measure μ_N over spin configurations:

$$\omega_{\beta,\Lambda}(f) = \frac{\mu_\Lambda(f e^{-\beta H_\Lambda})}{\mu_\Lambda(e^{-\beta H_\Lambda})}, \tag{5.1}$$

with

$$\mu_\Lambda(f) = \frac{1}{2^{|\Lambda|}} \sum_{\sigma \in \{-1,1\}^{|\Lambda|}} f(\sigma), \tag{5.2}$$

and f a smooth bounded function of the spin configurations. Such a deformed state $\omega_{\beta,\Lambda}$ fulfills a remarkable *stability property* with respect to further small deformations (perturbations): considering the Hamiltonian per particle

$$h(\sigma) = \frac{H_\Lambda(\sigma)}{|\Lambda|} \tag{5.3}$$

and the perturbation with parameter λ defined as

$$\omega^{(\lambda)}_{\beta,\Lambda}(f) = \frac{\omega_{\beta,\Lambda}(fe^{-\lambda h})}{\omega_{\beta,\Lambda}(e^{-\lambda h})}, \tag{5.4}$$

the Gibbs–Boltzmann measure is stable, i.e. λ-independent, in the thermodynamic limit $\Lambda \nearrow \mathbb{Z}^d$. In fact, one can observe that the perturbation amounts to a small temperature shift:

$$\omega^{(\lambda)}_{\beta,\Lambda}(f) = \frac{\mu_\Lambda(fe^{-\beta H_\Lambda - \lambda h})}{\mu_\Lambda(e^{-\beta H_\Lambda - \lambda h})} = \omega_{\beta+\frac{\lambda}{|\Lambda|},\Lambda}(f). \tag{5.5}$$

It is straightforward to see from the previous expression that when the expectations with respect to the Boltzmann–Gibbs measure are continuous with respect to the temperature (this might exclude points which are isolated singularities, possibly related to phase transitions) the perturbation has a vanishing effect in the large-volume limit. More precisely, one can prove the stability as follows.

Lemma 5.1 (Stability for deterministic systems) *For all inverse temperature intervals* $[\beta_0, \beta_1]$ *and for all bounded spin functions* f*, one has*

$$\lim_{\Lambda \nearrow \mathbb{Z}^d} \int_{\beta_0}^{\beta_1} [\omega_{\beta,\Lambda}(fh) - \omega_{\beta,\Lambda}(f)\omega_{\beta,\Lambda}(h)] d\beta = 0. \tag{5.6}$$

Proof Using (5.5) one has

$$\int_{\beta_0}^{\beta_1} \frac{d\omega^{(\lambda)}_{\beta,\Lambda}(f)}{d\lambda} d\beta = \frac{1}{|\Lambda|} \int_{\beta_0}^{\beta_1} \frac{d\omega^{(\lambda)}_{\beta,\Lambda}(f)}{d\beta} d\beta = \frac{\omega^{(\lambda)}_{\beta_1,\Lambda}(f) - \omega^{(\lambda)}_{\beta_0,\Lambda}(f)}{|\Lambda|} \tag{5.7}$$

from which one obtains (due to the boundedness of f)

$$\lim_{\Lambda \nearrow \mathbb{Z}^d} \int_{\beta_0}^{\beta_1} \frac{d\omega^{(\lambda)}_{\beta,\Lambda}(f)}{d\lambda} d\beta = 0 \qquad \forall\ \lambda \in \mathbb{R}, \quad \forall\ [\beta_0, \beta_1]. \tag{5.8}$$

As a consequence, computing the derivative at $\lambda = 0$, (5.6) is obtained. □

For the special case $f = h$, the previous lemma implies that the Hamiltonian per particle converges to a constant for large volumes with respect to the Gibbs measure, at least in β-integral average. Higher order derivatives of $\omega^{(\lambda)}_{\beta,\Lambda}(h)$ with respect to λ evaluated at $\lambda = 0$ (i.e. cumulants of h) are then compelled to vanish, since cumulants are homogeneous polynomials of the constant values of h with coefficients whose sum is zero.

Equation (5.6) has interesting consequences. It says, for instance, that the order parameter of a mean-field ferromagnetic Hamiltonian (the magnetization) has a trivial distribution in the thermodynamic limit (i.e. converges to a constant). In the Curie–Weiss model at zero magnetic field, for which the Hamiltonian per particle is the square of the magnetization, the previous identity implies that (by choosing $f = h$),

$$\omega_\beta(\sigma_1\sigma_2\sigma_3\sigma_4) = \omega_\beta(\sigma_1\sigma_2)^2 \tag{5.9}$$

in β-integral average (Contucci *et al.* (2002)). One can indeed prove that (5.9) holds for all β using the methods developed in Ellis and Newman (1978). The choice between $f = h^n$ and equivalently higher order derivatives in λ of the perturbed state, gives the well-known factorization property of the $2n$-point function as an nth power of the 2-point function (as discussed in Section 1.4).

In the case of disordered systems, such as spin glasses, the formulation of a stability property is more subtle, since a generic perturbation will affect both the temperature of the system and the disorder distribution. Moreover, the quenched state involves multiple copies and one could perturb any number of these copies. In the following we act on one copy, conventionally called 1, and study the consequences.

Definition 5.2 (Perturbed quenched state) For a spin glass model with Hamiltonian (1.8) and a bounded random spin function f on multiple copies, we define its expectation in the perturbed quenched state as

$$\langle\langle f\rangle\rangle^{(\lambda)}_{\beta,\Lambda} = \frac{\mathbb{E}\left[\Omega_{\beta,\Lambda}(f e^{-\lambda h(\sigma^{(1)})})\right]}{\mathbb{E}\left[\Omega_{\beta,\Lambda}(e^{-\lambda h(\sigma^{(1)})})\right]}. \tag{5.10}$$

Remark 5.3 We observe that this new perturbation is the analog (for the quenched measure of a random Hamiltonian) of the standard perturbation (5.4) introduced for deterministic systems with respect to the Boltzmann–Gibbs measure. On the other hand, we notice that while the standard perturbation for deterministic systems amounts to a small temperature shift, the newly introduced perturbation cannot be reduced by just a small temperature change, it also involves a small change in the disorder (see Remark 5.5). More precisely, the explicit expression of (5.10) reads

$$\langle\langle f \rangle\rangle_{\beta,\Lambda}^{(\lambda)} = \frac{\mathbb{E}\left[\frac{\sum_{\sigma^{(1)},\dots,\sigma^{(R)}} f(\sigma^{(1)},\dots,\sigma^{(R)}) e^{-\left(\beta+\frac{\lambda}{|\Lambda|}\right)H_\Lambda(\sigma^{(1)})} e^{-\beta\sum_{j=2}^{R} H_\Lambda(\sigma^{(j)})}}{\sum_{\sigma^{(1)},\dots,\sigma^{(R)}} e^{-\beta H_\Lambda(\sigma^{(1)})} e^{-\beta\sum_{j=2}^{R} H_\Lambda(\sigma^{(j)})}}\right]}{\mathbb{E}\left[\frac{\sum_\sigma e^{-\left(\beta+\frac{\lambda}{|\Lambda|}\right)H_\Lambda(\sigma)}}{\sum_\sigma e^{-\beta H_\Lambda(\sigma)}}\right]} \tag{5.11}$$

which clearly shows that only the numerator of the random Boltzmann–Gibbs state is affected by the change.

With the above definition, we can state the following stability property for disordered systems.

Theorem 5.4 (Spin glass stability and Ghirlanda–Guerra identities) *The quenched state of a Gaussian spin glass with Hamiltonian (1.8) is stable under the deformation (5.10), i.e. for all inverse temperature intervals* $[\beta_0, \beta_1]$

$$\lim_{\Lambda \nearrow \mathbb{Z}^d} \int_{\beta_0}^{\beta_1} \left. \frac{d\langle\langle f \rangle\rangle_{\beta,\Lambda}^{(\lambda)}}{d\lambda} \right|_{\lambda=0} d\beta = 0. \tag{5.12}$$

Moreover, the property (5.12) implies that, in β*-integral average, the whole set of so-called Ghirlanda–Guerra identities hold: for a bounded function* f *of the generalized overlaps* $\{c_{i,j}\}$ *(with* $i, j \in \{1, \dots, R\}$*)*

$$\lim_{\Lambda \nearrow \mathbb{Z}^d} \int_{\beta_0}^{\beta_1} \left[\langle f\, c_{1,R+1} \rangle_{\beta,\Lambda} - \frac{1}{R} \langle f \rangle_{\beta,\Lambda} \langle c_{1,2} \rangle_{\beta,\Lambda} - \frac{1}{R} \sum_{j=2}^{R} \langle f c_{1,j} \rangle_{\beta,\Lambda} \right] d\beta = 0. \tag{5.13}$$

Proof A simple calculation shows that

$$\left. \frac{d\langle\langle f \rangle\rangle_{\beta,\Lambda}^{(\lambda)}}{d\lambda} \right|_{\lambda=0} = \langle f h \rangle_{\beta,\Lambda} - \langle f \rangle_{\beta,\Lambda} \langle h \rangle_{\beta,\Lambda}. \tag{5.14}$$

The righthand side can be broken down into two terms which can be identified as the thermal and the disorder correlations between f and h:

$$\begin{aligned} \left. \frac{d\langle\langle f \rangle\rangle_{\beta,\Lambda}^{(\lambda)}}{d\lambda} \right|_{\lambda=0} &= \mathbb{E}\left[\Omega_{\beta,\Lambda}(fh) - \Omega_{\beta,\Lambda}(f)\Omega_{\beta,\Lambda}(h)\right] \\ &\quad + \mathbb{E}\left[\Omega_{\beta,\Lambda}(f)\Omega_{\beta,\Lambda}(h)\right] - \mathbb{E}\left[\Omega_{\beta,\Lambda}(f)\right]\mathbb{E}\left[\Omega_{\beta,\Lambda}(h)\right]. \end{aligned} \tag{5.15}$$

The two differences on the righthand side of the previous equation will be proved to converge to zero in β-integral average in Section 5.2 (Theorem 5.9) and in

Section 5.5 (Theorem 5.29), respectively. Moreover, in Theorems 5.10 and 5.30 it will be shown how the use of integration by parts yields Eq. (5.13). □

Remark 5.5 It is interesting to notice that the stability property (5.12) admits a simple formulation in terms of the cumulant generating function. Defining

$$\psi_{\beta,\Lambda}(\lambda) = \ln\langle e^{\lambda h}\rangle_{\beta,\Lambda} = \ln \mathbb{E}\left[\frac{\mathcal{Z}_{\beta+\lambda/|\Lambda|}}{\mathcal{Z}_{\beta}}\right], \tag{5.16}$$

then (5.12) with $R = 1$ (i.e. f is a function of only 1 copy) is equivalent, by the boundedness of f and the Schwarz inequality

$$|\langle hf\rangle - \langle h\rangle\langle f\rangle| \leqslant \sqrt{\langle f^2\rangle}\sqrt{\langle h^2\rangle - \langle h\rangle^2},$$

to the property of asymptotic flatness at the origin

$$\lim_{\Lambda\nearrow\mathbb{Z}^d} \int_{\beta_0}^{\beta_1} \left.\frac{d^2\psi_{\beta,\Lambda}(\lambda)}{d\lambda^2}\right|_{\lambda=0} d\beta = 0. \tag{5.17}$$

In particular, defining the generating function of thermal fluctuations as

$$\bar{\psi}_{\beta,\Lambda}(\lambda) = \mathbb{E}\left[\ln \omega_{\beta,\Lambda}(e^{\lambda h})\right] \tag{5.18}$$

and the generating function of disorder fluctuations as

$$\tilde{\psi}_{\beta,\Lambda}(\lambda) = \ln \mathbb{E}\left[e^{\lambda \omega_{\beta,\Lambda}(h)}\right], \tag{5.19}$$

one has

$$\frac{d^2\psi_{\beta,\Lambda}(\lambda)}{d\lambda^2} = \frac{d^2\bar{\psi}_{\beta,\Lambda}(\lambda)}{d\lambda^2} + \frac{d^2\tilde{\psi}_{\beta,\Lambda}(\lambda)}{d\lambda^2}. \tag{5.20}$$

Remark 5.6 The relevance of the stability properties and of the Ghirlanda–Guerra identities has been shown in Arguin (2008) and Panchenko (2010a). Under the hypothesis of discreteness of the overlap distribution, it was proved that competing particle systems satisfying invariance under reshuffling (Arguin) and spin systems satisfying Ghirlanda–Guerra identities (Panchenko) do fulfill the hierarchical structure (ultrametricity) originally introduced in the Parisi work for the mean-field spin glass (Mézard *et al.* (1987)).

5.2 Stochastic stability identities

In this section we review *stochastic stability* and show some of its consequences expressed in terms of the quenched equilibrium state – both in the form of identities for the overlap distribution, and of quenched additivity of the free energy. Stochastic stability provided a first and simple method to produce an infinite family of identities for the generalized overlap variables. Identities for random variables with respect to the quenched state reduce the degrees of freedom of the model and go toward the core of the Parisi mean-field theory: spin glasses are described by a probability distribution of a single overlap variable, and the joint overlaps distribution of more copies – necessary to describe the whole equilibrium state – can be obtained by a suitable combinatorial rule called *ultrametricity* which holds for classes of equivalent overlap structures (*overlap equivalence*). Although a similar research project is not yet completed for the Sherrington–Kirkpatrick model, important progress has been made toward it and there are clear indications, some based on numerical work (Contucci *et al.* (2006, 2007, 2009c)), some on rigorous grounds (Contucci (2003); Contucci and Giardinà (2005, 2007)), that mean-field models and short-range finite-dimensional models behave quite similarly as far as the factorization rules are concerned. Stochastic stability is also deeply rooted within theoretical physics. In fact it has been used in Franz *et al.* (1998, 1999) to determine a relation between off-equilibrium dynamics which are experimentally accessible and their static properties, and is considered (from the theoretical point of view) a structural property of the spin glass phase (see Parisi (2001, 2006)).

Definition 5.7 (Deformed quenched state) Given the Gaussian process $H_\Lambda(\sigma)$ of covariance $\mathcal{C}_\Lambda(\sigma,\tau)$ and the independent Gaussian process $K_\Lambda(\sigma)$ defined by the covariance $c_\Lambda(\sigma,\tau) = \mathcal{C}_\Lambda(\sigma,\tau)/|\Lambda|$, we introduce the deformed quenched expectation of a random bounded spin function f of multiple copies

$$\langle f\rangle^{(\lambda)}_{\beta,\Lambda} = \mathbb{E}\left[\frac{\Omega_{\beta,\Lambda}(f e^{\sqrt{\lambda}[K_\Lambda(\sigma^{(1)})+\cdots+K_\Lambda(\sigma^{(R)})]})}{\Omega_{\beta,\Lambda}(e^{\sqrt{\lambda}[K_\Lambda(\sigma^{(1)})+\cdots+K_\Lambda(\sigma^{(R)})]})}\right]. \tag{5.21}$$

Theorem 5.8 (Stochastic stability) *The spin glass quenched state is stochastically stable in β-average. Namely*

$$\lim_{\Lambda\nearrow\mathbb{Z}^d}\int_{\beta_0}^{\beta_1}\frac{d\langle f\rangle^{(\lambda)}_{\beta,\Lambda}}{d\lambda}d\beta = 0 \qquad \forall\,\lambda\in\mathbb{R},\quad \forall\ [\beta_0,\beta_1]. \tag{5.22}$$

Proof For $\widetilde{H}$ (independent from H and K) and distributed like H, we have, in distribution,

$$-\beta H_\Lambda + \sqrt{\lambda}K_\Lambda \stackrel{\mathcal{D}}{=} -\sqrt{\beta^2+\frac{\lambda}{|\Lambda|}}\,\widetilde{H}_\Lambda. \tag{5.23}$$

From the definition of the deformed quenched state of the function f (cf. Eq. (5.21)), the expectation $\langle f\rangle^{(\lambda)}_{\beta,\Lambda}$ turns out to be a function of $\beta^2+\frac{\lambda}{|\Lambda|}$. Hence we can write

$\langle f \rangle_{\beta,\Lambda}^{(\lambda)} = g(\beta^2 + \frac{\lambda}{|\Lambda|})$. Using the composite function derivation rule, we deduce:

$$\frac{d}{d\lambda}\langle f \rangle_{\beta,\Lambda}^{(\lambda)} = g'\left(\beta^2 + \frac{\lambda}{|\Lambda|}\right) \cdot \frac{1}{|\Lambda|} \tag{5.24}$$

where the prime denotes derivative w.r.t. the argument, and

$$\frac{d}{d\beta}\langle f \rangle_{\beta,\Lambda}^{(\lambda)} = g'\left(\beta^2 + \frac{\lambda}{|\Lambda|}\right) \cdot 2\beta, \tag{5.25}$$

from which we have

$$2\beta \frac{d}{d\lambda}\langle f \rangle_{\beta,\Lambda}^{(\lambda)} = \frac{1}{|\Lambda|}\frac{d}{d\beta}\langle f \rangle_{\beta,\Lambda}^{(\lambda)}. \tag{5.26}$$

Integrating over $d\beta$ and using the fundamental theorem of calculus, we obtain

$$\int_{\beta_0^2}^{\beta_1^2} \frac{d\langle f \rangle_{\beta,\Lambda}^{(\lambda)}}{d\lambda} d\beta^2 = \frac{\langle f \rangle_{\beta_1,\Lambda}^{(\lambda)} - \langle f \rangle_{\beta_0,\Lambda}^{(\lambda)}}{|\Lambda|}. \tag{5.27}$$

Remembering the assumption on boundedness of function f, one immediately obtains the proof in β-average. □

By a straightforward use of the Schwarz inequality, one can also show that a stochastic stability with respect to a deformation on a single copy holds in the spin glass phase. Namely, defining a single copy (say the first copy) deformed state as

$$\langle f \rangle_{\beta,\Lambda}^{(\lambda)_1} = \mathbb{E}\left[\frac{\Omega_{\beta,\Lambda}(f e^{\sqrt{\lambda}K_\Lambda(\sigma^{(1)})})}{\Omega_{\beta,\Lambda}(e^{\sqrt{\lambda}K_\Lambda(\sigma^{(1)})})}\right], \tag{5.28}$$

one has the following.

Theorem 5.9 (Stochastic stability, deformation 1 copy) *For all thermodynamically stable spin glass Gaussian models*

$$\lim_{\Lambda \nearrow \mathbb{Z}^d} \int_{\beta_0}^{\beta_1} \frac{d\langle f \rangle_{\beta,\Lambda}^{(\lambda)_1}}{d\lambda} d\beta = 0 \qquad \forall\ \lambda \in \mathbb{R}, \quad \forall\ [\beta_0, \beta_1]. \tag{5.29}$$

As a consequence

$$\lim_{\Lambda \nearrow \mathbb{Z}^d} \int_{\beta_0}^{\beta_1} \mathbb{E}\left[\Omega_{\beta,\Lambda}(f h(\sigma^{(1)})) - \Omega_{\beta,\Lambda}(f)\Omega_{\beta,\Lambda}(h(\sigma^{(1)}))\right] d\beta = 0. \tag{5.30}$$

Proof Using the equality in distribution (5.23), the computation of the derivative in

$$\int_{\beta_0}^{\beta_1} \frac{d\langle f \rangle_{\beta,\Lambda}^{(\lambda)_1}}{d\lambda} d\beta \tag{5.31}$$

gives

$$\int_{\beta_0}^{\beta_1} \mathbb{E}\left[\Omega^{(\lambda)_1}_{\beta,\Lambda}(fh(\sigma^{(1)})) - \Omega^{(\lambda)_1}_{\beta,\Lambda}(fh(\sigma^{(1)}))\Omega^{(\lambda)_1}_{\beta,\Lambda}(fh(\sigma^{(1)}))\right] \frac{d\beta}{2\sqrt{\beta^2 + \frac{\lambda}{|\Lambda|}}}. \tag{5.32}$$

Applying repeatedly the Schwarz inequality and using the boundedness of the function f, one finds that

$$\int_{\beta_0}^{\beta_1} \frac{d\langle f\rangle^{(\lambda)_1}_{\beta,\Lambda}}{d\lambda} d\beta \leqslant c\left(\beta_0, \beta_1, \frac{\lambda}{|\Lambda|}\right) \sqrt{\frac{\mathbb{E}\left[\omega^{(\lambda)_1}_{\beta_2,\Lambda}(h(\sigma^{(1)})) - \omega^{(\lambda)_1}_{\beta_1,\Lambda}(h(\sigma^{(1)}))\right]}{|\Lambda|}} \tag{5.33}$$

with $c(\beta_0, \beta_1, \frac{\lambda}{|\Lambda|})$ a bounded function with respect to the volume. Since the energy density expectations are bounded in the hypotheses of thermodynamically stable models (see Eq. (3.4)), the integral vanishes in the large-volume limit. Equation (5.30) follows by evaluating the derivative in Eq. (5.29) in $\lambda = 0$. □

The stochastic stability leads to a family of relations among moments in the form of zero-average polynomials (Aizenman and Contucci (1998)). These can be obtained as a special case of the following.

Theorem 5.10 (Stochastic stability identities) *Let f be a smooth bounded function defined on R copies. The following expression holds true (with $c_{l,l} = 1$):*

$$\lim_{\Lambda \nearrow \mathbb{Z}^d} \int_{\beta_0}^{\beta_1} d\beta \left\langle \sum_{k=2}^{R} f\, c_{1,k} - 2Rf\, c_{1,R+1} + (R+1) f\, c_{R+1,R+2} \right\rangle_{\Lambda,\beta} = 0. \tag{5.34}$$

Proof We evaluate separately the two terms in the lefthand side of Eq. (5.30) by using integration by parts for centered correlated Gaussian random variables:

$$\mathbb{E}\left[X_i \psi(X_1, \ldots, X_n)\right] = \sum_{j=1}^{n} \mathbb{E}\left[X_i X_j\right] \mathbb{E}\left[\partial_j \psi(X_1, \ldots, X_n)\right]. \tag{5.35}$$

It is convenient to denote by $p(R)$ the Gibbs–Boltzmann weight of R copies of the deformed system

$$p(R) = \frac{e^{-\beta\left[\sum_{k=1}^{R} H_\Lambda(\sigma^{(k)})\right]}}{[\mathcal{Z}(\beta)]^R}, \tag{5.36}$$

so that we have

$$-\frac{1}{\beta}\frac{dp(R)}{dH_\Lambda(\tau)} = p(R)\left(\sum_{k=1}^{R} \delta_{\sigma^{(k)},\tau}\right) - Rp(R)\frac{e^{-\beta[H_\Lambda(\tau)]}}{[\mathcal{Z}(\beta)]}. \tag{5.37}$$

We obtain

$$\begin{aligned}\mathbb{E}\left[\Omega(h(\sigma^{(1)})\, f)\right] &= \frac{1}{|\Lambda|}\,\mathbb{E}\left[\sum_{\sigma^{(1)},\ldots,\sigma^{(R)}} f\; H_\Lambda(\sigma^{(1)})\; p(R)\right] \\ &= \mathbb{E}\left[\sum_{\sigma^{(1)},\ldots,\sigma^{(R)}} \sum_{\tau} f\; c_\Lambda(\sigma^{(1)},\tau)\,\frac{dp(R)}{dH_\Lambda(\tau)}\right] \\ &= -\beta\left[\langle f\rangle + \sum_{k=2}^{R}\langle f\, c_{1,k}\rangle - R\langle f\, c_{1,\,R+1}\rangle\right]\end{aligned} \tag{5.38}$$

where we made use of the integration by parts formula. Analogously, the other term reads

$$\begin{aligned}\mathbb{E}\left[\Omega(h(\sigma^{(1)}))\,\Omega(f)\right] &= \frac{1}{|\Lambda|}\,\mathbb{E}\left[\sum_{\sigma^{(1)}}\sum_{\tau^{(1)},\ldots,\tau^{(R)}} f\; H_\Lambda(\sigma^{(1)})\; p(R+1)\right] \\ &= \mathbb{E}\left[\sum_{\sigma^{(1)}}\sum_{\tau^{(1)},\ldots,\tau^{(R)}}\sum_{\gamma} f\; c_\Lambda(\sigma^{(1)},\gamma)\,\frac{dp(R+1)}{dH_\Lambda(\gamma)}\right] \\ &= -\beta\left[\langle f\rangle + R\langle f\, c_{1,\,R+1}\rangle - (R+1)\langle f\, c_{R+1,\,R+2}\rangle\right].\end{aligned} \tag{5.39}$$

Inserting (5.38) and (5.39) in Eq. (5.30), the theorem is proved. □

5.3 Strong stochastic stability and marginal independence

In this section we show that a strong version of the stochastic stability property admits an equivalent formulation in terms of a new notion of independence for Gaussian fields.

Definition 5.11 (Marginal independence) Given any finite collection of identical and independent centered Gaussian fields $K_1(\sigma), K_2(\sigma), \ldots, K_l(\sigma)$ (independent also of the Hamiltonian) with covariance

$$\mathbb{E}\left[K_i(\sigma)K_j(\tau)\right] = \delta_{i,j}c_\Lambda(\sigma,\tau) \tag{5.40}$$

and any family of smooth, polynomially bounded, functions $F_1, F_2, \ldots, F_l$, a Gaussian spin glass model fulfills the property of asymptotic marginal independence if

$$\lim_{\Lambda \nearrow \mathbb{Z}^d} \mathbb{E}\left[\ln \omega_\Lambda \left(\exp\left(\sum_{i=1}^{l} F_i(K_i)\right)\right)\right] = \lim_{\Lambda \nearrow \mathbb{Z}^d} \sum_{i=1}^{l} \mathbb{E}\left[\ln \omega_\Lambda \left(\exp F_i(K_i)\right)\right]. \quad (5.41)$$

Remark 5.12 We observe that the previous notion of marginal independence (with respect to the ω_Λ measure) is a generalization of the usual notion of independence for standard random variables, which is the additivity of the cumulant generating function. The previous formula would, in fact, be trivial if the fields K_i would be mutually independent with respect to the measure ω_Λ, which is not the case here – since in general $\omega_\Lambda(F_i(K_i(\sigma))F_j(K_j(\sigma))) \neq \omega_\Lambda(F_i(K_i(\sigma)))\omega_\Lambda(F_j(K_j(\sigma)))$, even after the thermodynamic limit is taken.

Remark 5.13 Defining the cumulants of order p (for a given $p \in \mathbb{N}$) of the random variable g (distributed according to the Gibbs–Boltzmann measure) by

$$\omega(g; p) = \frac{\partial^p}{\partial \lambda^p} \ln \omega(e^{\lambda g})\big|_{\lambda=0}, \quad (5.42)$$

the marginal independence property implies, by simply computing derivatives from (5.41), the asymptotic linearity of the marginal cumulants, i.e.

$$\lim_{\Lambda \nearrow \mathbb{Z}^d} \mathbb{E}\left[\omega_\Lambda \left(\sum_{i=1}^{l} F_i(K_i); p\right)\right] = \lim_{\Lambda \nearrow \mathbb{Z}^d} \sum_{i=1}^{l} \mathbb{E}\left[\omega_\Lambda \left(F_i(K_i); p\right)\right]. \quad (5.43)$$

Definition 5.14 (Strong stochastic stability) Given the Gaussian process $H_\Lambda(\sigma)$ of covariance $\mathcal{C}_\Lambda(\sigma, \tau)$ and an independent Gaussian process $K_\Lambda(\sigma)$ defined by the covariance $c_\Lambda(\sigma, \tau)$, we introduce the strongly deformed quenched expectation of a smooth function f of independent fields $K_i(\sigma^{(j)})$ on possibly different copies

$$\langle f \rangle_{\beta,\Lambda}^{(F)} = \mathbb{E}\left[\frac{\Omega_{\beta,\Lambda}(f e^{F(K)})}{\Omega_{\beta,\Lambda}(e^{F(K)})}\right], \quad (5.44)$$

where K is one of the identically distributed K_i. A Gaussian spin glass model is stochastically stable in a strong sense if the deformed quenched state and the original one coincide in the thermodynamic limit:

$$\lim_{\Lambda \nearrow \mathbb{Z}^d} \langle f \rangle_{\beta,\Lambda}^{(F)} = \lim_{\Lambda \nearrow \mathbb{Z}^d} \langle f \rangle_{\beta,\Lambda}. \quad (5.45)$$

Theorem 5.15 *The properties of quenched independence (5.41) and strong stochastic stability (5.45) are equivalent.*

Proof The fact that (5.45) implies (5.41) is easily proved by iteration on a number of independent fields. The case of two fields is proved by observing that

$$\mathbb{E}\left[\ln \frac{\omega_\Lambda\left(e^{F_1(K_1)+F_2(K_2)}\right)}{\omega_\Lambda(e^{F_1(K_1)})}\right] = \mathbb{E}\left[\ln \omega_\Lambda^{(F_1)}\left(e^{F_2(K_2)}\right)\right] \tag{5.46}$$

and using strong stochastic stability. We observe, in fact, that the entire set of powers of the Boltzmann state necessary to compute the logarithm is obtained by transforming powers into computations on multiple copies. For example,

$$\mathbb{E}\left[\omega_\Lambda(K)^2\right] = \mathbb{E}\left[\omega_\Lambda(K(\sigma))\omega_\Lambda(K(\tau))\right] = \mathbb{E}\left[\Omega_\Lambda(K(\sigma)K(\tau))\right] = \langle c_{1,2}\rangle_\Lambda. \tag{5.47}$$

The other implication, namely (5.41) implies (5.45), is seen by observing that defining

$$\varphi_\Lambda(\varepsilon) = \mathbb{E}\left[\ln \frac{\omega_\Lambda(e^{\varepsilon G+F})}{\omega_\Lambda(e^{\varepsilon G})\omega_\Lambda(e^F)}\right], \tag{5.48}$$

(5.41) implies that for all ε

$$\lim_{\Lambda \nearrow \mathbb{Z}^d} \varphi_\Lambda(\varepsilon) = 0. \tag{5.49}$$

Since we can interchange the computation of the derivatives w.r.t. ε with the thermodynamical limit (see Simon (1993)), we have that the marginal independence property (5.41) is equivalent to the vanishing of all the quantities $\frac{d^n\varphi_\Lambda}{d\varepsilon^n}(\varepsilon)|_{\varepsilon=0}$ in the infinite-volume limit. We first observe that

$$\frac{d\varphi_\Lambda}{d\varepsilon}\Big|_{\varepsilon=0} = \mathbb{E}\left[\omega_\Lambda^{(F)}(G)\right] - \mathbb{E}\left[\omega_\Lambda(G)\right] \to 0, \tag{5.50}$$

which is the stability for the first power of the Gibbs–Boltzmann measure (one replica). The second derivative gives

$$\frac{d^2\varphi_\Lambda}{d\varepsilon}\Big|_{\varepsilon=0} = \mathbb{E}\left[\left(\omega_\Lambda^{(F)}(G)\right)^2\right] - \mathbb{E}\left[\omega_\Lambda^{(F)}(G^2)\right] - \mathbb{E}\left[(\omega_\Lambda(G))^2\right] + \mathbb{E}\left[\omega_\Lambda(G^2)\right] \to 0. \tag{5.51}$$

Thus

$$\lim_{\Lambda \nearrow \mathbb{Z}^d} \mathbb{E}\left[\left(\omega_\Lambda^{(F)}(G)\right)^2\right] - \mathbb{E}\left[(\omega_\Lambda(G))^2\right] = \lim_{\Lambda \nearrow \mathbb{Z}^d} \mathbb{E}\left[\omega_\Lambda^{(F)}(G^2)\right] - \mathbb{E}\left[\omega_\Lambda(G^2)\right] = 0 \tag{5.52}$$

where the last equality is applied to the smooth function G^2 by the first order equation (5.50). The proof of (5.45) follows by induction. □

We conclude this paragraph with two observations which lead to the following two lemmata. The first is that the strong stochastic stability for linear perturbations

λK is a natural consequence of a hypothesis of continuity in the temperature. From (5.23) one immediately derives that for all bounded smooth f

$$\langle f \rangle_{\Lambda,\beta}^{(\lambda K)} = \langle f \rangle_{\Lambda,\sqrt{\beta^2+\frac{\lambda^2}{|\Lambda|}}}. \tag{5.53}$$

The previous relation entails the following.

Lemma 5.16 (Continuity in temperature and stochastic stability) *If, in a certain temperature range, the quenched averages are uniformly continuous in β, as $\Lambda \nearrow \mathbb{Z}^d$, and the infinite-volume limits exist, then the quenched state is stable under linear deformations, i.e.:*

$$\langle f \rangle_{\beta}^{(\lambda K)} = \langle f \rangle_{\beta}. \tag{5.54}$$

The second observation is that the continuity assumptions made above also entail a peculiar stationarity property (related to the cavity approach) when applied to the Sherrington–Kirkpatrick model. Adding one spin to the system and tracing over it, one can easily see that

$$\langle f \rangle_{N+1,\beta} = \langle f \rangle_{N,\bar{\beta}}^{(\ln\cosh 2\bar{\beta}h)}, \tag{5.55}$$

where $\bar{\beta} = \sqrt{\frac{N}{N+1}}\beta$ and $h(\sigma)$ is a Gaussian field with covariance $q_N(\sigma,\sigma')$.

Lemma 5.17 *Under the temperature continuity assumption in the thermodynamical limit, the Sherrington–Kirkpatrick models fulfills the stability property:*

$$\langle f \rangle_{\beta} = \langle f \rangle_{\beta}^{(\ln\cosh 2\beta h)}. \tag{5.56}$$

5.4 Graph theoretical approach to Aizenman–Contucci identities

When generic observables of the previous section are specialized to polynomials in the covariance entries, the stochastic stability theory identifies an infinite family of polynomials with zero average with respect to the quenched measure. Those polynomials admit a graph-theoretical description, which was introduced in Aizenman and Contucci (1998) and further developed in Bianchi *et al.* (2004), where a complete classification was provided. The origin of the graph-theoretical approach is the Wick rule of the Gaussian integration which, for the quenched measure where the disorder and thermal averages are intertwined, takes the form that we describe in this section.

We are going to use the notion of edge-labeled multigraphs (see, for instance, Diestel (2000)).

Definition 5.18 (Multigraphs and overalp algebra) A multigraph (on $R \in \mathbb{N}$ vertices) is a finite set E and a map

$$G : E \to [V]^2 \cup V \tag{5.57}$$

with the vertex set $V = \{1, \ldots, R\}$ and the family $[V]^2 := \{\{i, j\} \mid i \neq j \in V\}$ of unordered vertex pairs. We call the elements of $E' := G^{-1}(V)$ *legs*, and the elements of $E'' := G^{-1}([V]^2) = E - E'$ *edges*. Let $\mathcal{G}$ be the set of all multigraphs, i.e.

$$\mathcal{G} = \bigcup_{m,n=0}^{\infty} \mathcal{G}^{(m,n)} \quad \text{with} \quad \mathcal{G}^{(m,n)} := \{G \mid |E''| = m, |E'| = n\}. \tag{5.58}$$

In the setting above, multiple edges and legs are allowed and their multiplicities are indicated as exponents. For example, the graph

$$G = \{1, 2\}\{2, 3\}^2\{2\}^2\{3\} \in \mathcal{G}^{(3,3)}$$

has edges $\{1, 2\}$ of multiplicity 1, edges $\{2, 3\}$ of multiplicity 2, leg $\{2\}$ of multiplicity 2 and leg $\{3\}$ of multiplicity 1. We indicate by $\mathcal{G}'' := \bigcup_m \mathcal{G}^{(m,0)}$ the subset of multigraphs without legs, and $\mathcal{G}' := \bigcup_n \mathcal{G}^{(0,n)}$ the subset of multigraphs without edges. We also observe that $\mathcal{G}$ is the basis of the vector space $\widetilde{\mathcal{G}} := \mathbb{C}[\mathcal{G}]$ of finite linear combinations, and we denote subspaces spanned by the various subsets of $\mathcal{G}$ by a tilde (using multiplication of multigraphs and formal linear combinations, $\widetilde{\mathcal{G}}$ becomes an algebra, sometimes called *overlap algebra*). We now introduce the following.

Definition 5.19 (Wick contraction) The linear operator $\mathcal{C} : \widetilde{\mathcal{G}} \to \widetilde{\mathcal{G}}$, called *Wick contraction*, is defined by

$$\mathcal{C}(E) := E'' \, \mathcal{C}\left(E'\right), \quad \mathcal{C}\left(E'\right) := \sum_{\text{pairings } \pi} \prod_{i=1}^{m} \{G(\pi(2i-1)), G(\pi(2i))\}, \tag{5.59}$$

for the decomposition $E = E' \cup E''$ of the multigraph $G \in \mathcal{G}$ (here, for $|E'| = 2m$, the pairings are seen as permutations $\pi : \{1, \ldots, 2m\} \to E' \cong \{1, \ldots, 2m\}$). Note that

$$\mathcal{C}(\widetilde{\mathcal{G}}^{(m,2l+1)}) = \{0\} \quad \text{and} \quad \mathcal{C}(\widetilde{\mathcal{G}}^{(m,2l)}) \subseteq \widetilde{\mathcal{G}}^{(m+l,0)}. \tag{5.60}$$

Remark 5.20 For a family of independent centered Gaussian fields $\{k_i\}$ with covariance $\mathbb{E}\left[k_i k_j\right] = c_{i,j}$, the previous definition of a Wick contraction for graphs is motivated by the bijection $I : \hat{G} \mapsto G$ between overlap monomials

$$\hat{G} := \left(\prod_{1 \le i < j \le R} c_{i,j}^{m_{i,j}} \right) \left(\prod_{i=1}^{R} k_i^{n_i} \right) \tag{5.61}$$

and multigraphs $G : E \to [V]^2 \cup V$, with edge multiplicities $|G^{-1}(i, j)| = m_{i,j}$ and leg multiplicities $|G^{-1}(i)| = n_i$, since

$$I\, \mathbb{E}\left[\hat{G}\right] = \mathcal{C} I(\hat{G}). \tag{5.62}$$

This allows us to use the symbols G and $\hat{G}$ in the sequel indifferently.

In order to study the properties of the deformed quenched measure with the graph-theoretical approach, we need the following:

Definition 5.21 (Gaussian derivation operator) The linear operator $\delta : \widetilde{\mathcal{G}} \to \widetilde{\mathcal{G}}$ is defined by

$$\delta G := \sum_{v \in V(G)} \delta_v G, \tag{5.63}$$

with

$$\delta_v := \delta_v^{(+)} + \delta_v^{(-)}, \quad \text{and} \quad \delta_v^{(+)} G := \{v\} G, \quad \delta_v^{(-)} G := -\{v'\} G, \tag{5.64}$$

where v' is the first element of $\mathbb{N}$ not belonging to V (so $v' = R + 1$ if $V = \{1, \ldots, R\}$). Note that

$$\delta \left(\widetilde{\mathcal{G}}^{(m,l)} \right) \subseteq \widetilde{\mathcal{G}}^{(m,l+1)} \tag{5.65}$$

and

$$\delta(G_1 G_2) = (\delta G_1) G_2 + G_1 \delta G_2, \quad \delta(\emptyset) = 0 \tag{5.66}$$

are the properties of a *derivation* ($\emptyset$ denoting the graph without legs and edges, the neutral element under composition of multigraphs).

Here are a few examples: $\delta_2^{(+)}\{1, 2\} = \{2\}\{1, 2\}$, $\delta_3^{(-)}\{1, 3\} = -\{2\}\{1, 3\}$, $\delta_3^{(-)}\delta_3^{(-)}\{1, 3\} = \{4\}\{2\}\{1, 3\}$.

We will now illustrate in graph-theoretical terms the consequences of the stochastic stability property for the couple $(\langle - \rangle, C)$, where C is the random matrix with elements $c_{i,j}$ (recall Definition 1.12). To study those consequences, we will have to consider derivatives of expectations of G with respect to suitable parameters: our computations will tacitly use the fact that the thermodynamic limit does commute with the operation of derivatives as was proved, for every order of derivation, in the appendix of Aizenman and Contucci (1998).

The invariance under permutations of indexes of the replicas (see Definition 1.13) allows us to use the multigraph isomorphisms induced by arbitrary relabeling of the vertices freely, in order to calculate expectations.

Recalling Definition 5.14 of the strong stochastic stability property, Equation (5.45) implies that

$$\langle G\rangle^{(\lambda)} = \langle G\rangle \quad \text{for all } G. \tag{5.67}$$

Therefore, as the lefthand side is λ-independent

$$\frac{d^n}{d\lambda^n}\langle G\rangle^{(\lambda)} = 0. \tag{5.68}$$

We now show the relation between the standard notion of derivative of the deformed state and the graph-theoretical operator δ.

Theorem 5.22 (Derivative of the deformed state and the operator δ)

$$\frac{d}{d\lambda}\langle G\rangle^{(\lambda^2)} = \langle \delta G\rangle. \tag{5.69}$$

Proof Since the deformed state is computed in $\sqrt{\lambda^2} = \lambda$, the action of the operator δ on multigraphs corresponds to the usual derivative with respect to the parameter λ. Such a derivative produces a *truncated correlation* expressed in the rule (5.64). The derivation property (5.66) represents the Leibniz rule

$$\frac{d}{d\lambda}\Omega^{(\lambda^2)}(-) = \sum_{l=1}^{R}\left(\frac{d}{d\lambda}\omega_l^{(\lambda^2)}(-)\right)\prod_{r\neq l}\omega_r^{(\lambda^2)}(-) \tag{5.70}$$

for the derivative on the replica space. Each differentiation with respect to λ produces a sum of monomials, each containing an additional zero mean Gaussian variable K. □

Consequently, for every $G \in \widetilde{\mathcal{G}}''$ and every integer n,

$$\frac{d^n}{d\lambda^n}\langle G\rangle^{(\lambda^2)} = \langle \delta^n G\rangle^{(\lambda^2)}. \tag{5.71}$$

When computed in $\lambda = 0$, the fields K produced by the operator δ are independent from those that appear at the exponential, and can be contracted using the Wick rule:

$$\frac{d^n}{d\lambda^n}\langle G\rangle^{(\lambda^2)}|_{\lambda=0} = \langle \delta^n G\rangle^{(\lambda^2)}|_{\lambda=0} = \langle C\delta^n G\rangle. \tag{5.72}$$

For odd values of n, the previous expectation is zero for parity reasons, while for even values of n, the set of identities (cf. Theorem 5.10)

$$\langle \mathcal{C}\delta^{2n} G \rangle = 0 \tag{5.73}$$

is a consequence of stochastic stability.

Example 5.23 Let us explicitly see the first nontrivial case of the previous identities obtained from $G = \{1, 2\}$. One finds:

$$\mathcal{C}\delta^2\{1, 2\} = 2\left[\{1, 2\}^2 - 4\{1, 2\}\{2, 3\} + 3\{1, 2\}\{3, 4\}\right]. \tag{5.74}$$

The following two theorems show that the entire set of identities that hold for every G and every n is already included in the case $n = 1$.

Theorem 5.24 (Derivative of the deformed state and the operator Δ) *Defining the operator Δ as*

$$\frac{d}{d\lambda}\langle G \rangle^{(\lambda)} = \frac{1}{2}\langle \Delta G \rangle^{(\lambda)}, \tag{5.75}$$

we have

$$\Delta = \mathcal{C}\delta^2. \tag{5.76}$$

Proof A direct computation of the first derivative gives

$$\frac{d}{d\lambda}\langle G \rangle^{(\lambda^2)} = 2\lambda \frac{d}{d\lambda^2}\langle G \rangle^{(\lambda^2)} = \lambda \langle \Delta G \rangle^{(\lambda^2)}. \tag{5.77}$$

As for the second derivative, one has

$$\frac{d^2}{d\lambda^2}\langle G \rangle^{(\lambda^2)} = \langle \Delta G \rangle^{(\lambda^2)} + \lambda^2 \langle \Delta^2 G \rangle^{(\lambda^2)}. \tag{5.78}$$

Computing the previous equation in $\lambda = 0$ one has by (5.72) for every $G \in \widetilde{\mathcal{G}}''$

$$\langle \mathcal{C}\delta^2 G \rangle = \langle \Delta G \rangle. \tag{5.79}$$

□

Theorem 5.25 (Relation between the operators δ and Δ)

$$\mathcal{C}\delta^{2n} = (2n-1)!!\, \Delta^n \qquad (n \in \mathbb{N}_0), \tag{5.80}$$

or equivalently, in the sense of a formal power series,

$$\mathcal{C}\exp(t\delta) = \exp\left(\tfrac{1}{2}\Delta t^2\right) \qquad (t \in \mathbb{C}), \tag{5.81}$$

which justifies the name Gaussian *derivation for the operator δ.*

Proof From the Leibniz rule, one has the formula (valid for all f)

$$\frac{d^n(xf)}{dx^n} = n\frac{d^{n-1}f}{dx^{n-1}} + x\frac{d^n f}{dx^n}. \tag{5.82}$$

Now, as in the computation of the first derivative (5.77), one has

$$\frac{d^k}{d\lambda^k}\langle G\rangle^{(\lambda^2)} = \frac{d^{k-1}}{d\lambda^{k-1}}[\lambda\langle \Delta G\rangle^{(\lambda^2)}]. \tag{5.83}$$

Using formula (5.82), one finds

$$\frac{d^k}{d\lambda^k}\langle G\rangle^{(\lambda^2)} = (k-1)\frac{d^{k-2}}{d\lambda^{k-2}}\langle \Delta G\rangle^{(\lambda^2)} + \lambda\frac{d^{k-1}}{d\lambda^{k-1}}\langle \Delta G\rangle^{(\lambda^2)}. \tag{5.84}$$

Computing the previous identity in $\lambda = 0$ for $k = 2n$

$$\frac{d^{2n}}{d\lambda^{2n}}\langle G\rangle^{(\lambda^2)}\bigg|_{\lambda=0} = (2n-1)\frac{d^{2n-2}}{d\lambda^{2n-2}}\langle \Delta G\rangle^{(\lambda^2)}\bigg|_{\lambda=0} \tag{5.85}$$

which is equivalent to

$$\mathcal{C}\delta^{2n} = (2n-1)\mathcal{C}\delta^{2n-2}\Delta, \tag{5.86}$$

and, iterating, the claim of the theorem follows. □

Remark 5.26 In order to fully appreciate the content of (5.80), we examine the case $n = 2$:

$$\mathcal{C}\delta^4 = 3\Delta^2 \tag{5.87}$$

(see also Contucci (2002)). It is interesting to note that the lefthand side of (5.87) contains *a priori* $3 \cdot 2^4 = 48$ terms (the factor 3 coming from the Wick contraction of a fourth order monomial and 2^4 being the number of terms of $(\delta^+ + \delta^-)^4$). The righthand side contains only $4^2 = 16$ terms, coming from the square of Δ. Although, by definition, the Wick contraction does not conserve the number of edges or vertices, the presence of alternating signs in the definition of δ, together with the invariance under permutation of graph labelings, produces a delicate cancellation mechanism responsible for the clean equality (5.87).

To conclude the section we give a combinatorial characterization of the overlap polynomials of the form ΔG. The main result is a formula which allows the computation of the polynomials ΔG from a quadratic expression in the number R of *real*-replicas, evaluated at $R = 0$. In the original replica symmetry breaking formulation, a set of identities was obtained in the limit $n \to 0$ under the so-called hypothesis of *replica equivalence* (see Mézard *et al.* (1987); Parisi (2004); Parisi and Ricci-Tersenghi (2000)), which says that in the $n \times n$ Parisi matrix Q, each row is a permutation of any other. We stress however that the two methods, although producing a similar set of identities, are conceptually different: one works with the real-replicas matrix $C = \{c_{l,m}\}$ and deduces identities from stochastic stability, while the other deals with the replica symmetry breaking matrix $Q = \{Q_{a,b}\}$ (whose indexes do not label real replicas, but label the "dual" variables introduced in (1.58)) and obtains the identities from the assumption of replica equivalence.

Theorem 5.27 (Stochastic stability identities and zero replicas) *Define*

$$M_R = \sum_{i \neq j=1}^{R} c_{i,j} \tag{5.88}$$

for all integers $R \geqslant 1$. For all G and all $R > |G| + 1$, the quantity $\langle G M_R \rangle$ is a quadratic function of R. Its polynomial extension to all real R fulfills the following formula

$$\langle \Delta G \rangle = \langle G M_R \rangle|_{R=0} = 0. \tag{5.89}$$

Example 5.28 To illustrate the statement, let us take $G = c_{1,2}$. In this case, the lefthand side of (5.89) is given by (5.74) and

$$\langle c_{1,2} M_R \rangle = 2\langle c_{1,2}^2 \rangle + 4(R-2)\langle c_{1,2} c_{2,3} \rangle + (R-2)(R-3)\langle c_{1,2} c_{3,4} \rangle. \tag{5.90}$$

The two coincide at $R = 0$ (defined by polynomial extension).

Proof The proof proceeds through the explicit computation of the lefthand and righthand sides of (5.89). First, we notice that using the definition of Δ for $|G| = l$

$$\Delta G = \sum_{v \neq v'} G \cdot \{v, v'\} - 2l \sum_{v} G \cdot \{v, \tilde{v}\} + l(l+1) G \cdot \{\tilde{v}, \tilde{v}'\}, \tag{5.91}$$

where v and v' are summed over the set of vertices of G, and $\tilde{v}$ and $\tilde{v}'$ denote a pair of external vertices. Moreover, we observe that the invariance under permutation of the quenched measure implies that

$$\begin{aligned} \langle G M_R \rangle = &\sum_{v \neq v'} \langle G \cdot \{v, v'\} \rangle + 2(R-l) \sum_{v} \langle G \cdot \{v, \tilde{v}\} \rangle \\ &+ (R-l)(R-l-1) \langle G \cdot \{\tilde{v}, \tilde{v}'\} \rangle. \end{aligned} \tag{5.92}$$

When computed in $R = 0$, (5.92) coincides with (5.91) and they both vanish as a consequence of the stochastic stability property. □

The method of stochastic stability and the identities that can be derived from it have been further studied in Barra (2006); Sollich and Barra (2012); Arguin (2008); Panchenko (2010a, b, 2011b).

5.5 Identities from self-averaging

In this section we investigate the property of vanishing of the fluctuations of the intensive quantities with respect to the disorder. In particular, the property of self-averaging of the random internal energy per particle will lead to a family of identities which, together with those coming from stochastic stability, provide the whole set of Ghirlanda–Guerra identities.

Theorem 5.29 (Internal energy self-averaging) *The random internal energy is a self-averaging quantity, in β-average. Namely, for all thermodynamically stable spin glass Gaussian models, one has*

$$\lim_{\Lambda \nearrow \mathbb{Z}^d} \int_{\beta_0}^{\beta_1} \left(\mathbb{E}\left[\omega^2_{\beta,\Lambda}(h)\right] - \mathbb{E}\left[\omega_{\beta,\Lambda}(h)\right]^2 \right) d\beta = 0 \qquad \forall\, [\beta_0, \beta_1]. \tag{5.93}$$

As a consequence, the following identity holds for every bounded function f defined on R copies:

$$\lim_{\Lambda \nearrow \mathbb{Z}^d} \int_{\beta_0}^{\beta_1} \left(\mathbb{E}\left[\Omega_{\beta,\Lambda}(f)\Omega_{\beta,\Lambda}(h)\right] - \mathbb{E}\left[\Omega_{\beta,\Lambda}(f)\right]\mathbb{E}\left[\Omega_{\beta,\Lambda}(h)\right] \right) d\beta = 0. \tag{5.94}$$

Proof The result is obtained in two steps which use general theorems of measure theory. First we obtain the convergence to zero almost everywhere in β of the variance of the internal energy from Theorem 3.10, then we apply the Lebesgue dominated convergence theorem to obtain convergence to zero in β-average of the same quantity.

We know from Chapter 3 that the sequence of convex functions given by the random pressure per particle $\mathcal{P}_\Lambda(\beta)/|\Lambda|$ converges to a non-random limiting value $p(\beta)$, which coincides with the limit of its average. The limit is finite because of the thermodynamic limit stability assumption. The convergence is in L_2, i.e. the variance vanishes as seen in Eq. (3.41) of Theorem 3.10. By general convexity arguments (Ruelle (1999)), it follows that the sequence of the derivatives $\omega_{\beta,\Lambda}(h) = \mathcal{P}'_\Lambda(\beta)/|\Lambda|$ converges to $u(\beta) = p'(\beta)$ almost everywhere in β. In fact the vanishing of the variance of a sequence of convex functions is inherited in all points in which the derivative exists (which is almost everywhere for a convex function) by the sequence of its derivatives. The convergence is in L_2, hence we conclude that the random internal energy is a self-averaging quantity almost everywhere in the temperature:

$$\lim_{\Lambda \nearrow \mathbb{Z}^d} \left(\mathbb{E}\left[\omega^2_{\beta,\Lambda}(h)\right] - \mathbb{E}\left[\omega_{\beta,\Lambda}(h)\right]^2 \right) = 0 \quad \text{almost everywhere in } \beta. \tag{5.95}$$

In order to obtain the convergence in β-average we use the Lebesgue dominated convergence theorem. In fact we prove that the sequence of variances of $\omega_{\beta,\Lambda}(h)$

is uniformly bounded (in every interval $[\beta_0, \beta_1]$) by an integrable function of β. A lengthy but simple computation, which uses integration by parts, gives

$$\mathbb{E}\left[\omega_{\beta,\Lambda}(H_\Lambda)\right] = \mathbb{E}\left[\sum_{X\subset\Lambda} J_X\omega(\sigma_X)\right] = \sum_{X\subset\Lambda}\beta\Delta_X^2\left[1-\mathbb{E}\left[\omega^2(\sigma_X)\right]\right] \leqslant \beta|\Lambda|\bar{c} \tag{5.96}$$

$$\begin{aligned}\mathbb{E}\left[\omega^2_{\beta,\Lambda}(H_\Lambda)\right] &= \mathbb{E}\left[\sum_{X,Y\subset\Lambda} J_X J_Y\omega(\sigma_X)\omega(\sigma_Y)\right]\\ &= \sum_{X,Y\subset\Lambda}\beta^2\Delta_X^2\Delta_Y^2\,\mathbb{E}\left[1-\omega^2(\sigma_X)-\omega^2(\sigma_Y)+6\omega^2(\sigma_X)\omega^2(\sigma_Y)\right.\\ &\qquad\left.-\,6\omega(\sigma_X)\omega(\sigma_Y)\omega(\sigma_X\sigma_Y)+\omega^2(\sigma_X\sigma_Y)\right] \qquad (5.97)\\ &\leqslant 14\beta^2|\Lambda|^2\bar{c}^2 \qquad (5.98)\end{aligned}$$

from which, using the thermodynamic stability assumption (Eq. (3.4))

$$\mathbb{E}\left[\omega^2_{\beta,\Lambda}(h)\right] - \mathbb{E}\left[\omega_{\beta,\Lambda}(h)\right]^2 \leqslant 15\beta^2\bar{c}^2. \tag{5.99}$$

Thic concludes the proof of (5.93). Eq. (5.94) is then obtained by a simple application of the Schwartz inequality, using the assumption on the boundedness of f. □

Theorem 5.30 (Identities from self-averaging) *Let f be a smooth bounded function defined on R copies. The following expression holds true*

$$\lim_{\Lambda\nearrow\mathbb{Z}^d}\int_{\beta_0}^{\beta_1} d\beta\left[R\langle f c_{1,R+1}\rangle_{\beta,\Lambda} - (R+1)\langle f c_{R+1,R+2}\rangle_{\beta,\Lambda} + \langle f\rangle_{\beta,\Lambda}\langle c_{1,2}\rangle_{\beta,\Lambda}\right] = 0. \tag{5.100}$$

Proof The identities follow by explicit evaluation of the two terms in (5.94) using integration by parts. Actually, since the first term was already evaluated in (5.39), we are left with the explicit evaluation of the second term which simply gives

$$\begin{aligned}\mathbb{E}\left[\Omega(h(\sigma^{(1)}))\right]\mathbb{E}\left[\Omega(f)\right] &= \frac{1}{|\Lambda|}\mathbb{E}\left[\sum_{\sigma^{(1)}} H_\Lambda(\sigma^{(1)})\,p_\Lambda(1)\right]\langle f\rangle\\ &= \mathbb{E}\left[\sum_{\sigma^{(1)}}\sum_{\gamma} c_\Lambda(\sigma^{(1)},\gamma)\,\frac{dp_\Lambda(1)}{dH_\Lambda(\gamma)}\right]\langle f\rangle\\ &= -\beta\,\langle f\rangle_{\beta,\Lambda}[1-\langle c_{1,2}\rangle_{\beta,\Lambda}]. \qquad (5.101)\end{aligned}$$

Inserting (5.39) and (5.101) into Eq. (5.94), we obtain (5.100). □

Remark 5.31 Putting together the Aizenman–Contucci stochastic stability identities proved in Theorem 5.10 and those proved in Theorem 5.30 from the internal energy self-averaging, we obtain the entire set of Ghirlanda–Guerra identities stated in Theorem 5.4.

The results proved here are based on the hypothesis of thermodynamic stability and do not take into account the fine structure of each considered model. In the case of the Sherrington–Kirkpatrick model, one can however prove that the Ghirlanda–Guerra identities hold in distribution (see Talagrand (2010b)).

Moreover, while we limit ourselves to a statement in β-average for expectations of bounded functions of R copies, in the works Chatterjee (2009), Arguin and Chatterjee (2010), and Panchenko (2010b), the Ghirlanda–Guerra identities have been proven to hold at every β at which the free energy is differentiable.

5.6 Identities on the Nishimori line

In this section we show that for a general spin glass model with non-centered couplings there exists a family of identities involving expectations of generalized overlaps and generalized magnetizations in the quenched state. To obtain the Ghirlanda–Guerra identities we show that the generalized magnetization per particle is a non-fluctuating quantity for large volumes (the result holds in μ-average, where μ is a suitable parameter that tunes the overall average interaction). Finally, thanks to the special properties of the spin glass systems on the Nishimori line, we show that the identities hold pointwise and are reached at the rate of the inverse volume.

We now need to extend the notion of algebra of observables, since in the non-zero average interaction case, the relevant quantities are not only expectations of the overlap but also of the general magnetization.

Definition 5.32 (Generalized magnetization random variables) Considering a Gaussian Hamiltonian

$$H_\Lambda(\sigma) := \sum_{X\subset\Lambda} J_X\sigma_X, \tag{5.102}$$

with $\mathbb{E}[J_X] = \mu_X$ and $\mathbb{E}\left[J_X^2\right] = \Delta_X^2$ we define

$$m_\Lambda(\sigma) = \frac{1}{|\Lambda|}\sum_{X\subset\Lambda} \mu_X\sigma_X, \tag{5.103}$$

$$c_\Lambda(\sigma,\tau) = \frac{1}{|\Lambda|}\sum_{X\subset\Lambda} \Delta_X^2\sigma_X\tau_X, \tag{5.104}$$

and, for any smooth bounded function $G(m_\Lambda, c_\Lambda)$ of the mean and covariance matrix entries, we introduce the random R-dimensional vector of elements $\{m_k\}$ (called *generalized magnetization*) and the $R \times R$ matrix of elements $\{c_{k,l}\}$ by the formula

$$\langle G(m,c)\rangle_{\beta,\Lambda} := \mathbb{E}\left[\Omega_{\beta,\Lambda}(G(m_\Lambda, c_\Lambda))\right]. \tag{5.105}$$

Theorem 5.33 (Identities for non-centered interactions) *For a Hamiltonian fulfilling the conditions of thermodynamic stability, such that there exists a constant $\bar{c}$ such that*

$$\sup_{\Lambda\subset\mathbb{Z}^d} \frac{1}{|\Lambda|}\sum_{X\subset\Lambda} |\mu_X| \leqslant \bar{c} < \infty$$

$$\sup_{\Lambda\subset\mathbb{Z}^d} \frac{1}{|\Lambda|}\sum_{X\subset\Lambda} \Delta_X^2 \leqslant \bar{c} < \infty, \tag{5.106}$$

defining

$$f_\Lambda^{(1)}(\beta,\mu) = \sum_{l=1}^{R} \langle (m_l - m_{R+1})\, G\rangle_{\beta,\Lambda} +$$

$$-\beta\left(\left\langle \sum_{\substack{k,l=1\\k\neq l}}^{R} G\, c_{l,k} - 2RG\sum_{l=1}^{R} c_{l,\,R+1} + R(R+1)G\, c_{R+1,\,R+2}\right\rangle_{\beta,\Lambda}\right) \tag{5.107}$$

and

$$f_\Lambda^{(2)}(\beta,\mu) = \langle m_{R+1} G\rangle_{\beta,\Lambda} - \langle m_1\rangle_{\beta,\Lambda}\langle G\rangle_{\beta,\Lambda} +$$

$$-\beta\left(\sum_{k=1}^{R+1}\langle G\, c_{k,\,R+1}\rangle_{\beta,\Lambda} - (R+1)\langle G\, c_{R+1,\,R+2}\rangle_{\beta,\Lambda} - \langle G\rangle_{\beta,\Lambda}(1-\langle c_{1,2}\rangle_{\beta,\Lambda})\right), \tag{5.108}$$

the quenched equilibrium state fulfills, for every observable G and every temperature interval $[\beta_1,\beta_2]$, the following identities in the thermodynamic limit

$$\lim_{\Lambda\nearrow\mathbb{Z}^d} \int_{\beta_1}^{\beta_2} d\beta\, |f_\Lambda^{(1)}(\beta,\mu)| = 0 \tag{5.109}$$

$$\lim_{\Lambda\nearrow\mathbb{Z}^d} \int_{\beta_1}^{\beta_2} d\beta\, |f_\Lambda^{(2)}(\beta,\mu)| = 0. \tag{5.110}$$

Proof The proof follows the standard strategy of the internal energy fluctuations control: the thermal fluctuations, related to stochastic stability (Section 5.2), and those with respect to the disorder (Section 5.5), which produce the identities involving $f^{(1)}$ and $f^{(2)}$ respectively after integration by parts. The differences that emerge in comparison to the Ghirlanda–Guerra identities are related to the presence of the generalized magnetization random variables which come from the non-zero average interactions. The bound on the fluctuations are based on the thermodynamic stability conditions (5.106). For further details, see Contucci *et al.* (2009b). □

Theorem 5.34 (Ghirlanda–Guerra identities for non-centered interactions) *Under the same hypotheses of the previous theorem, and defining*

$$g_{\Lambda}^{(1)}(\beta,\mu) = -\beta\left(\left\langle \sum_{\substack{k,l=1\\k\neq l}}^{R} G\,c_{l,k} - 2RG\sum_{l=1}^{R} c_{l,\,R+1} + R(R+1)G\,c_{R+1,\,R+2}\right\rangle_{\beta,\,\Lambda}\right) \tag{5.111}$$

and

$$g_{\Lambda}^{(2)}(\beta,\mu) = -\beta\left(\sum_{k=1}^{R+1}\langle Gc_{k,\,R+1}\rangle_{\beta,\Lambda} - (R+1)\langle Gc_{R+1,\,R+2}\rangle_{\beta,\Lambda} - \langle G\rangle_{\beta,\Lambda}(1-\langle c_{1,2}\rangle_{\beta,\Lambda})\right), \tag{5.112}$$

introducing the random variables J'_X and the parameter μ as

$$J_X = J'_X + \mu_X, \tag{5.113}$$

$$\mu\mu'_X = \mu_X, \tag{5.114}$$

such that

$$J_X = J'_X + \mu\mu'_X, \tag{5.115}$$

the quenched equilibrium state of a thermodynamically stable Hamiltonian fulfills, for every observable G and every set $[\beta_1,\beta_2]\times[\mu_1,\mu_2]$, the following identities in the thermodynamic limit

$$\lim_{\Lambda\nearrow\mathbb{Z}^d}\int_{\mu_1}^{\mu_2} d\mu \int_{\beta_1}^{\beta_2} d\beta\,|g_{\Lambda}^{(1)}(\beta,\mu)| = 0 \tag{5.116}$$

$$\lim_{\Lambda\nearrow\mathbb{Z}^d}\int_{\mu_1}^{\mu_2} d\mu \int_{\beta_1}^{\beta_2} d\beta\,|g_{\Lambda}^{(2)}(\beta,\mu)| = 0. \tag{5.117}$$

Proof The proof follows directly from Theorem 5.33 and from the fact that for every interval $[\mu_1, \mu_2]$, in the thermodynamic limit

$$\lim_{\Lambda \nearrow \mathbb{Z}^d} \int_{\mu_1}^{\mu_2} d\mu (\langle m^2 \rangle_{\beta,\Lambda} - \langle m \rangle^2_{\beta,\Lambda}) = 0. \tag{5.118}$$

The vanishing of the generalized magnetization fluctuations can be achieved from the same strategy that has been pursued so far to control the fluctuation of the internal energy. One has, in fact,

$$\begin{aligned} \langle m^2 \rangle_{\beta,\Lambda} - \langle m \rangle^2_{\beta,\Lambda} &= \mathbb{E}\left[\Omega_{\beta,\Lambda}(m_\Lambda(\sigma)^2) - (\Omega_{\beta,\Lambda}(m_\Lambda(\sigma)))^2\right] \\ &\quad + \mathbb{E}\left[(\Omega_{\beta,\Lambda}(m_\Lambda(\sigma)))^2\right] - \left(\mathbb{E}\left[\Omega_{\beta,\Lambda}(m_\Lambda(\sigma))\right]\right)^2 \end{aligned} \tag{5.119}$$

and we observe that generalized magnetization is related to the pressure by

$$\frac{\partial}{\partial \mu}\left(\frac{\mathcal{P}_\Lambda}{|\Lambda|}\right) = \frac{\beta}{\mu}\frac{1}{|\Lambda|}\Omega_{\beta,\Lambda}\left(\sum_{X \in \Lambda} \mu_X \sigma_X\right) = \frac{\beta}{\mu}\,\Omega_{\beta,\Lambda}(m_\Lambda(\sigma)) \tag{5.120}$$

The fluctuations w.r.t. the Gibbs state (first term on the righthand side of Eq. (5.119)) are easily controlled by a stochastic stability argument:

$$\begin{aligned} &\int_{\mu_1}^{\mu_2} d\mu \mathbb{E}\left[\Omega_{\beta,\Lambda}(m_\Lambda(\sigma)^2) - (\Omega_{\beta,\Lambda}(m_\Lambda(\sigma)))^2\right] \\ &= \frac{1}{|\Lambda|}\int_{\mu_1}^{\mu_2} d\mu\, \frac{\mu^2}{\beta^2}\,\mathbb{E}\left[\frac{\partial}{\partial \mu}\frac{\beta}{\mu}\Omega_{\beta,\Lambda}(m_\Lambda(\sigma))\right], \end{aligned} \tag{5.121}$$

where the righthand side can be bounded, integrating by parts in μ, by $\beta^{-1}3(\mu_2 - \mu_1)$.

The fluctuations w.r.t. the disorder (second term on the righthand side of Eq. (5.119)) are bounded by the same argument of Theorem 5.29. Indeed, from self-averaging of the pressure per particle

$$\mathbb{E}\left[(p_\Lambda^2)\right] - (\mathbb{E}\left[(p_\Lambda)\right])^2 \leqslant \frac{\bar{c}}{|\Lambda|} \to 0, \tag{5.122}$$

and convexity of the finite-volume pressure

$$\frac{\partial^2}{\partial \mu^2}\frac{\mathcal{P}_\Lambda}{|\Lambda|} = |\Lambda|\frac{\beta^2}{\gamma^2}[\Omega_{\beta,\Lambda}(m_\Lambda(\sigma)^2) - (\Omega_{\beta,\Lambda}(m_\Lambda(\sigma)))^2] \geqslant 0,$$

one also deduces that the sequence of derivatives (5.120) is self-averaging in μ-average. Hence, in the limit of large volumes,

$$\int_{\mu_1}^{\mu_2} d\mu \left[\mathbb{E}\left[(\Omega_{\beta,\Lambda}(m_\Lambda(\sigma)))^2\right] - \left(\mathbb{E}\left[\Omega_{\beta,\Lambda}(m_\Lambda(\sigma))\right]\right)^2\right] \to 0. \tag{5.123}$$

□

Theorem 5.35 (Pointwise identities on the Nishimori line) *In the space of parameters $(\mu_X, \Delta_X^2)_{X\in\Lambda}$ that defines the Nishimori line*

$$\mu_X = \lambda_X \Delta_X^2 \tag{5.124}$$

the identities of Theorem 5.33 hold pointwise, namely

$$\lim_{\Lambda\nearrow\mathbb{Z}^d} \tilde{f}_\Lambda^{(1)}(\gamma) = 0 \tag{5.125}$$

$$\lim_{\Lambda\nearrow\mathbb{Z}^d} \tilde{f}_\Lambda^{(2)}(\gamma) = 0 \tag{5.126}$$

where the quantities $\tilde{f}^{(i)}$, defined by the righthand side of (5.107) and (5.108) computed on the Nishimori line, depend only on the parameters

$$\gamma_X = \frac{\mu_X^2}{\Delta_X^2}. \tag{5.127}$$

Proof The strategy here is to prove the vanishing of fluctuations of the Hamiltonian per particle $h(\sigma)$. This implies, following the same reasoning as in previous sections, the vanishing of correlations with a generic bounded function G using the Schwarz inequality:

$$|\langle hG\rangle_\Lambda - \langle h\rangle_\Lambda \langle G\rangle_\Lambda| \leqslant \sqrt{\langle h^2\rangle_\Lambda - \langle h\rangle_\Lambda^2}\sqrt{\langle G^2\rangle_\Lambda - \langle G\rangle_\Lambda^2}. \tag{5.128}$$

On the Nishimori manifold, the random internal energy per particle is self-averaging. More precisely, the following result holds:

$$\langle h^2\rangle_\Lambda - \langle h\rangle_\Lambda^2 \leqslant \frac{\bar{c}}{|\Lambda|}. \tag{5.129}$$

This result is based on the explicit evaluation of the two terms on the lefthand side. The term $\langle h\rangle_\Lambda$ has been evaluated in Lemma 2.16 and gives:

$$\mathbb{E}\left[\Omega_\Lambda(H_\Lambda(\sigma))\right] = \sum_{X\subset\Lambda} \mu_X. \tag{5.130}$$

Similarly, the term $\langle h^2\rangle_\Lambda$ is evaluated:

$$\begin{aligned}
\mathbb{E}\left[\Omega_\Lambda(H_\Lambda(\sigma)^2)\right] &= \mathbb{E}\left[\Omega_\Lambda\left(\sum_{X,Y} J_X J_Y \sigma_X \sigma_Y\right)\right] \\
&= \mathbb{E}\left[\sum_{X,Y} J_X J_Y\right] \\
&= \sum_{X\neq Y} \mu_X\mu_Y + \sum_X (\mu_X^2 + \Delta_X^2) \\
&= \sum_{X,Y} \mu_X\mu_Y + \sum_X \Delta_X^2.
\end{aligned} \tag{5.131}$$

Therefore, using (5.130) and (5.131) in the expression for the variance of h, one finds

$$\begin{aligned} \langle h(\sigma)^2\rangle_\Lambda - \langle h(\sigma)\rangle_\Lambda^2 &= \frac{1}{|\Lambda|^2}\mathbb{E}\left[\Omega_\Lambda(H_\Lambda(\sigma)^2)\right] - \frac{1}{|\Lambda|^2}(\mathbb{E}\left[\Omega_\Lambda(H_\Lambda(\sigma))\right])^2 \\ &= \frac{1}{|\Lambda|^2}\sum_{X\subset\Lambda}\Delta_X^2 \leqslant \frac{\bar{c}}{|\Lambda|}. \end{aligned} \tag{5.132}$$

□

5.7 Interaction flip identities

We have seen in previous sections how the control of fluctuations of intensive thermodynamic quantities, like the internal energy per particle, leads to identities in the quenched spin glass state. In this section, we extend the method to other intensive thermodynamic functions and prove the vanishing of their fluctuations. We consider here the free energies of two spin glass systems that differ for having some set of interactions flipped ($J \to -J$). We show that their difference has a variance whose growth is bounded by the volume of the flipped region. To obtain the identities, we introduce a new interpolation method which extends the standard interpolation technique to the entire circle. Using integration by parts, the identities are expressed in terms of overlap moments. As a side result, the case of the non-interacting random field is analyzed and the triviality of its overlap distribution proved.

The interest in investigating these quantities stems also from the classical analysis of the domain wall stability in phase transition phenomena. As an example, one may consider the result that is stated in Newman and Stein (1992) (quoted there as proved by M. Aizenman and D. S. Fisher) for the difference (ΔF) between the free energy of the Edwards–Anderson model on a d-dimensional lattice with linear size L, and the volume L^d when going from periodic to antiperiodic boundary conditions on the hyperplane which is orthogonal to (say) the x-direction. This result is a bound for the variance of ΔF which grows no more than the volume of the hyperplane. Such an upper bound is equivalent to a bound for the stiffness exponent $\theta \leqslant (d-1)/2$ (see the discussion in Southern and Young (1977); Bray and Moore (1984); Fisher and Huse (1988); van Enter (1990)). Although that bound is not expected to be saturated, we prove here that it implies an identity for the equilibrium quantities. When expressed in terms of spin variables, some of the overlap identities that we find generalize the structure of truncated correlation functions that appear in Temesvari (2007).

Let us introduce some definitions which are needed in the sequel.

Definition 5.36 (Thermodynamic quantities in a flipped region) For a spin glass Hamiltonian with Gaussian couplings in a volume $\Lambda \subset \mathbb{Z}^d$,

$$H_\Lambda(\sigma) = -\sum_{X \subset \Lambda} J_X \sigma_X. \tag{5.133}$$

For a given subset $\Lambda' \subseteq \Lambda$, we write

$$H_\Lambda(\sigma) = H_{\Lambda'}(\sigma) + H_{\Lambda \setminus \Lambda'}(\sigma) \tag{5.134}$$

where

$$H_{\Lambda'}(\sigma) = -\sum_{X \subset \Lambda'} J_X \sigma_X, \quad H_{\Lambda \setminus \Lambda'}(\sigma) = -\sum_{\substack{X \subset \Lambda \\ X \not\subset \Lambda'}} J_X \sigma_X, \tag{5.135}$$

and

$$H_{\Lambda,\Lambda'}(\sigma) = -H_{\Lambda'}(\sigma) + H_{\Lambda \setminus \Lambda'}(\sigma) \tag{5.136}$$

denotes the Hamiltonian with the J couplings inside the region Λ' that have been flipped. We associate the natural thermodynamic quantities with this Hamiltonian as follows: the random flipped partition function

$$\mathcal{Z}_{\Lambda,\Lambda'}(\beta) := \sum_{\sigma \in \Sigma_\Lambda} e^{-\beta H_{\Lambda,\Lambda'}(\sigma)} \equiv \sum_{\sigma \in \Sigma_\Lambda} e^{\beta H_{\Lambda'}(\sigma) - \beta H_{\Lambda \setminus \Lambda'}(\sigma)}; \tag{5.137}$$

the random flipped free energy/pressure

$$-\beta \mathcal{F}_{\Lambda,\Lambda'}(\beta) := \mathcal{P}_{\Lambda,\Lambda'}(\beta) := \ln \mathcal{Z}_{\Lambda,\Lambda'}(\beta); \tag{5.138}$$

the quenched flipped free energy/pressure

$$-\beta F_{\Lambda,\Lambda'}(\beta) := P_{\Lambda,\Lambda'}(\beta) := \mathbb{E}\left[\mathcal{P}_{\Lambda,\Lambda'}(\beta)\right]. \tag{5.139}$$

Definition 5.37 (Interpolation on the circle) Let $\xi = \{\xi_i\}_{1 \le i \le n}$ and $\eta = \{\eta_i\}_{1 \le i \le n}$ be two independent families of centered Gaussian random variables, each having covariance matrix $\mathcal{C}$, i.e.

$$\begin{aligned} \mathbb{E}\left[\xi_i \xi_j\right] &= \mathcal{C}_{i,j} \\ \mathbb{E}\left[\eta_i \eta_j\right] &= \mathcal{C}_{i,j} \\ \mathbb{E}\left[\xi_i \eta_j\right] &= 0. \end{aligned} \tag{5.140}$$

The interpolating process $x(t) = \{x_i(t)\}_{1 \le i \le n}$, defined by

$$x_i(t) = \cos(t)\, \xi_i + \sin(t)\, \eta_i \tag{5.141}$$

is a stationary Gaussian process with covariance

$$\mathbb{E}\left[x_i(t)x_j(t)\right] = C_{i,j}\cos(t-s). \tag{5.142}$$

We introduce the interpolating random pressure:

$$\mathcal{P}(t) = \ln \mathcal{Z}(t) = \ln \sum_{i=1}^{n} e^{x_i(t)}, \tag{5.143}$$

and the notation $\langle C_{1,2}\rangle_{t,s}$ to denote the expectation of the covariance matrix in the deformed quenched state, constructed from two independent copies with Boltzmann weights $x(t)$ and $x(s)$. Hence:

$$\langle C_{1,2}\rangle_{t,s} = \mathbb{E}\left[\sum_{i,j=1}^{n} C_{i,j}\frac{e^{x_i(t)+x_j(s)}}{\mathcal{Z}(t)\mathcal{Z}(s)}\right]. \tag{5.144}$$

The definition is extended in the obvious way to more than two copies. The main object of our investigation is the random variable given by the difference between the pressures evaluated at the boundaries of the interval $[a, b]$

$$\mathcal{X}(a,b) = \mathcal{P}(b) - \mathcal{P}(a). \tag{5.145}$$

The following lemma gives an explicit expression for the first two moments of this random variable.

Lemma 5.38 (Quenched moments of pressure difference) *For the random variable $\mathcal{X}(a,b)$ defined above, we have*

$$\mathbb{E}\left[\mathcal{X}(a,b)\right] = 0 \tag{5.146}$$

and

$$\mathbb{E}\left[(\mathcal{X}(a,b))^2\right] = \int_a^b\int_a^b dtds\, k_1(t,s)\langle C_{1,2}\rangle_{t,s} - \int_a^b\int_a^b dt\, ds\, k_2(t,s)$$
$$\times\left[\left\langle C_{1,2}^2\right\rangle_{t,s} - 2\langle C_{1,2}C_{2,3}\rangle_{s,t,s} + \langle C_{1,2}C_{3,4}\rangle_{t,s,s,t}\right] \tag{5.147}$$

with

$$k_1(t,s) = \cos(t-s), \qquad k_2(t,s) = \sin^2(t-s). \tag{5.148}$$

Proof The stationarity of the Gaussian process $x(t)$ implies that $\mathbb{E}\left[\mathcal{P}(t)\right]$ is independent of t; this proves (5.146). As far as the computation of the second moment is concerned, starting from

$$\mathcal{X}(a,b) = \int_a^b dt \mathcal{P}'(t) = \int_a^b dt \sum_{i=1}^n x_i'(t) \frac{e^{x_i(t)}}{\mathcal{Z}(t)} \tag{5.149}$$

we have

$$\begin{aligned} \mathbb{E}\left[(\mathcal{X}(a,b))^2\right] &= \int_a^b dt \int_a^b ds \mathbb{E}\left[\mathcal{P}'(t)\mathcal{P}'(s)\right] \\ &= \int_a^b dt \int_a^b ds \sum_{i,j=1}^n \mathbb{E}\left[x_i'(t)x_j'(s)\frac{e^{x_i(t)+x_j(s)}}{\mathcal{Z}(t)\mathcal{Z}(s)}\right]. \end{aligned} \tag{5.150}$$

Using integration by parts, the computation of the average in the rightmost term of the previous formula gives

$$\begin{aligned} \mathbb{E}\left[x_i'(t)x_j'(s)\frac{e^{x_i(t)+x_j(s)}}{\mathcal{Z}(t)\mathcal{Z}(s)}\right] &= \cos(t-s)\langle C_{1,2}\rangle_{t,s} \\ - \sin^2(s-t)\Big(\big\langle C_{12}^2\big\rangle_{t,s} &- 2\langle C_{12}C_{23}\rangle_{t,s,t} + \langle C_{12}C_{34}\rangle_{t,s,s,t}\Big), \end{aligned} \tag{5.151}$$

proving the lemma. □

In the following results, we will prove a bound for the fluctuations of the quantity

$$\mathcal{X}_{\Lambda,\Lambda'} = \mathcal{P}_\Lambda - \mathcal{P}_{\Lambda,\Lambda'} \tag{5.152}$$

in the same spirit of a similar bound for the pressure given in Theorem 3.10. Our formulation applies to the general Gaussian spin glass, thus including both mean-field and finite-dimensional models.

Before stating the result it is useful to note that, as a consequence of the symmetry of the Gaussian distribution, the variation of the random pressure has a zero average:

$$\mathbb{E}\left[\mathcal{X}_{\Lambda,\Lambda'}\right] = 0. \tag{5.153}$$

Lemma 5.39 (Concentration of pressure difference) *For every subset $\Lambda' \subset \Lambda$, the disorder fluctuation of the free energy variation $\mathcal{X}_{\Lambda,\Lambda'}$ satisfies the following inequality: for all $x > 0$*

$$\mathbb{P}\left(|\mathcal{X}_{\Lambda,\Lambda'}| \geqslant x\right) \leqslant 2\exp\left(-\frac{x^2}{8\pi\beta^2\bar{c}|\Lambda'|}\right) \tag{5.154}$$

where $\bar{c}$ is the constant in the thermodynamic stability condition

$$\sup_{\Lambda \subset \mathbb{Z}^d} \frac{1}{|\Lambda|} \sum_{X \subset \Lambda} \Delta_X^2 \leqslant \bar{c} < \infty. \tag{5.155}$$

The variance of the free energy variation satisfies the bound

$$\mathrm{Var}(\mathcal{X}_{\Lambda,\Lambda'}) = \mathbb{E}\left[\mathcal{X}_{\Lambda,\Lambda'}^2\right] \leqslant 16\pi\,\bar{c}\,\beta^2\,|\Lambda'|. \tag{5.156}$$

Proof Consider an $s > 0$ and let $x > 0$. By Markov's inequality, one has

$$\begin{aligned}\mathbb{P}\left\{\mathcal{X}_{\Lambda,\Lambda'} \geqslant x\right\} &= \mathbb{P}\left\{\exp[s\mathcal{X}_{\Lambda,\Lambda'}] \geqslant \exp(sx)\right\} \\ &\leqslant \mathbb{E}\left[\exp[s\mathcal{X}_{\Lambda,\Lambda'}]\right]\exp(-sx).\end{aligned} \tag{5.157}$$

To bound the generating function

$$\mathbb{E}\left[\exp[s\mathcal{X}_{\Lambda,\Lambda'}]\right] \tag{5.158}$$

one introduces, for a parameter $t \in [0, \frac{\pi}{2}]$, the following interpolating partition functions:

$$\mathcal{Z}^+(t) = \sum_{\sigma \in \Sigma_\Lambda} e^{-\beta \cos t\, H_{\Lambda'}^{(1)}(\sigma) - \beta H_{\Lambda\setminus\Lambda'}^{(3)}(\sigma) - \beta \sin t\, H_{\Lambda'}^{(2)}(\sigma)}, \tag{5.159}$$

$$\mathcal{Z}^-(t) = \sum_{\sigma \in \Sigma_\Lambda} e^{\beta \cos t\, H_{\Lambda'}^{(1)}(\sigma) - \beta H_{\Lambda\setminus\Lambda'}^{(3)}(\sigma) + \beta \sin t\, H_{\Lambda'}^{(2)}(\sigma)}. \tag{5.160}$$

Here the Hamiltonians $H_{\Lambda'}^{(1)}(\sigma)$, $H_{\Lambda'}^{(2)}(\sigma)$, $H_{\Lambda\setminus\Lambda'}^{(3)}(\sigma)$, defined in (5.135), depend on three independent copies $\{J_X^{(1)}\}_{X\subset\Lambda}$, $\{J_X^{(2)}\}_{X\subset\Lambda}$, $\{J_X^{(3)}\}_{X\subset\Lambda}$ of the centered Gaussian disorder with variance $\mathbb{E}\left[J_X^2\right] = \Delta_X^2$. Now we are ready to consider the interpolating function

$$\phi(t) = \ln \mathbb{E}_3\left[\mathbb{E}_1\left[\exp\left(s\,\mathbb{E}_2\left[\ln\frac{\mathcal{Z}^+(t)}{\mathcal{Z}^-(t)}\right]\right)\right]\right], \tag{5.161}$$

where $\mathbb{E}_1[-]$, $\mathbb{E}_2[-]$ and $\mathbb{E}_3[-]$ denote expectation with respect to the three independent families of Gaussian variables J_X. The following may be verified immediately:

$$\phi(0) = \ln \mathbb{E}\left[\exp[s\,\mathcal{X}_{\Lambda,\Lambda'}]\right], \tag{5.162}$$

and, using (5.153),

$$\phi\left(\frac{\pi}{2}\right) = 0. \tag{5.163}$$

This implies that

$$\mathbb{E}\left[\exp[s\mathcal{X}_{\Lambda,\Lambda'}]\right] = e^{\phi(0)-\phi(\frac{\pi}{2})} = e^{-\int_0^{\frac{\pi}{2}} \phi'(t)dt}. \tag{5.164}$$

On the other hand, the function $\phi'(t)$ can be easily bounded. Defining

$$K(t) = \exp\left(s\mathbb{E}_2\left[\ln\frac{\mathcal{Z}^+(t)}{\mathcal{Z}^-(t)}\right]\right), \tag{5.165}$$

the derivative is given by

$$\phi'(t) = \phi'_+(t) + \phi'_-(t) \tag{5.166}$$

where

$$\phi'_\pm(t) = \frac{s\mathbb{E}_3\left[\mathbb{E}_1\left[K(t)\mathbb{E}_2\left[\frac{\mathcal{Z}^\pm(t)'}{\mathcal{Z}^\pm(t)}\right]\right]\right]}{\mathbb{E}_3\left[\mathbb{E}_1\left[K(t)\right]\right]}. \tag{5.167}$$

The first term in the derivative is

$$\phi'_+(t) = \frac{s\mathbb{E}_3\left[\mathbb{E}_1\left[K(t)\mathbb{E}_2\left[\sum_{\sigma\in\Sigma_\Lambda} p_t^+(\sigma)\left[\beta\sin t\, H^{(1)}_{\Lambda'}(\sigma) - \beta\cos t\, H^{(2)}_{\Lambda'}(\sigma)\right]\right]\right]\right]}{\mathbb{E}_3\left[\mathbb{E}_1\left[K(t)\right]\right]} \tag{5.168}$$

where

$$p_t^+(\sigma) = \frac{e^{-\beta\cos t\, H^{(1)}_{\Lambda'}(\sigma) - \beta H^{(3)}_{\Lambda\backslash\Lambda'}(\sigma) - \beta\sin t\, H^{(2)}_{\Lambda'}(\sigma)}}{\mathcal{Z}^+(t)}. \tag{5.169}$$

Applying the integration by parts formula, a simple computation gives

$$\begin{aligned}
&\beta\sin t\,\mathbb{E}_3\left[\mathbb{E}_1\left[K(t)\,\mathbb{E}_2\left[\sum_\sigma p_t^+(\sigma)H^{(1)}_{\Lambda'}(\sigma)\right]\right]\right]\\
&\quad= -s\beta^2\sin t\,\cos t\,\mathbb{E}_3\left[\mathbb{E}_1\left[K(t)\sum_{X\subset\Lambda'}\Delta_X^2[s_t^+(X)^2 + s_t^+(X)s_t^-(X)]\right]\right]\\
&\quad\quad-\beta^2\sin t\,\cos t\,\mathbb{E}_3\left[\mathbb{E}_1\left[K(t)\,\mathbb{E}_2\left[\sum_\sigma \mathcal{C}_{\Lambda'}(\sigma,\sigma)p_t^+(\sigma)\right]\right]\right]\\
&\quad\quad+\beta^2\sin t\,\cos t\,\mathbb{E}_3\left[\mathbb{E}_1\left[K(t)\,\mathbb{E}_2\left[\sum_{\sigma,\tau} \mathcal{C}_{\Lambda'}(\sigma,\tau)p_t^+(\sigma)p_t^+(\tau)\right]\right]\right]
\end{aligned} \tag{5.170}$$

and

$$-\beta \cos t \,\mathbb{E}_3\left[\mathbb{E}_1\left[K(t)\,\mathbb{E}_2\left[\sum_\sigma p_t^+(\sigma)\, H_{\Lambda'}^{(2)}(\sigma)\right]\right]\right]$$

$$= \beta^2 \sin t\, \cos t\, \mathbb{E}_3\left[\mathbb{E}_1\left[K(t)\,\mathbb{E}_2\left[\sum_\sigma \mathcal{C}_{\Lambda'}(\sigma,\sigma) p_t^+(\sigma)\right]\right]\right]$$

$$- \beta^2 \sin t\, \cos t\, \mathbb{E}_3\left[\mathbb{E}_1\left[K(t)\,\mathbb{E}_2\left[\sum_{\sigma,\tau} \mathcal{C}_{\Lambda'}(\sigma,\tau) p_t^+(\sigma) p_t^+(\tau)\right]\right]\right] \tag{5.171}$$

where

$$s_t^+(X) = \mathbb{E}_2\left[\sum_\sigma \sigma_X\, p_t^+(\sigma)\right], \qquad s_t^-(X) = \mathbb{E}_2\left[\sum_\sigma \sigma_X\, p_t^-(\sigma)\right] \tag{5.172}$$

and

$$p_t^-(\sigma) = \frac{e^{\beta \cos t\, H_{\Lambda'}^{(1)}(\sigma) - \beta H_{\Lambda\setminus\Lambda'}^{(3)}(\sigma) + \beta \sin t\, H_{\Lambda'}^{(2)}(\sigma)}}{\mathcal{Z}^-(t)}. \tag{5.173}$$

Taking the difference between (5.170) and (5.171), one finds that

$$\phi'_+(t) = -s^2\beta^2 \sin t\, \cos t\, \frac{\mathbb{E}_3\left[\mathbb{E}_1\left[K(t)\sum_{X\subset\Lambda'} \Delta_X^2 [s_t^+(X)^2 + s_t^+(X) s_t^-(X)]\right]\right]}{\mathbb{E}_3\left[\mathbb{E}_1\left[K(t)\right]\right]}. \tag{5.174}$$

With a similar computation, one also obtains

$$\phi'_-(t) = -s^2\beta^2 \sin t\, \cos t\, \frac{\mathbb{E}_3\left[\mathbb{E}_1\left[K(t)\sum_{X\subset\Lambda'} \Delta_X^2 [s_t^-(X)^2 + s_t^+(X) s_t^-(X)]\right]\right]}{\mathbb{E}_3\left[\mathbb{E}_1\left[K(t)\right]\right]}, \tag{5.175}$$

then we conclude that

$$\phi'(t) = -s^2\beta^2 \sin t\, \cos t\, \frac{\mathbb{E}_3\left[\mathbb{E}_1\left[K(t)\sum_{X\subset\Lambda'} \Delta_X^2 [s_t^+(X) + s_t^-(X)]^2\right]\right]}{\mathbb{E}_3\left[\mathbb{E}_1\left[K(t)\right]\right]}. \tag{5.176}$$

Using the thermodynamic stability condition (5.155), this yields

$$|\phi'(t)| \leqslant 4\,\beta^2\,\bar{c}\, s^2\, |\Lambda'| \tag{5.177}$$

from which it follows, using (5.164),

$$\mathbb{E}\left[\exp[s\mathcal{X}_{\Lambda,\Lambda'}]\right] \leqslant \exp\left(2\pi\beta^2\,\bar{c}\,s^2\,|\Lambda'|\right). \tag{5.178}$$

Inserting this bound into the inequality (5.157) and optimizing over s, one finally obtains

$$\mathbb{P}\left(\mathcal{X}_{\Lambda,\Lambda'} \geqslant x\right) \leqslant \exp\left(-\frac{x^2}{8\pi\,\beta^2\,\bar{c}\,|\Lambda'|}\right). \tag{5.179}$$

The proof of inequality (5.154) is completed by observing that one can repeat a similar computation for $\mathbb{P}(\mathcal{X}_{\Lambda,\Lambda'} \leqslant -x)$. The result for the variance (5.156) is then immediately proved, thanks to (5.153), using the identity

$$\mathbb{E}\left[\mathcal{X}^2_{\Lambda,\Lambda'}\right] = 2\int_0^{\infty} x\ \mathbb{P}(|\mathcal{X}_{\Lambda,\Lambda'}| \geqslant x)dx. \tag{5.180}$$

□

The main result of this section now follows.

Theorem 5.40 (Overlap identities from the difference of pressure) *Given a volume Λ, consider the Gaussian spin glass with Hamiltonian (2.18). For a subvolume $\Lambda' \subseteq \Lambda$ and a parameter $t \in [0, \pi]$, let*

$$\omega_t(-) = \sum_{\sigma}(-)\frac{e^{-H_\sigma(t)}}{\mathcal{Z}(t)} \tag{5.181}$$

with

$$H_\sigma(t) = \cos(t)H^{(1)}_{\Lambda'}(\sigma) + \sin(t)H^{(2)}_{\Lambda'}(\sigma) + H_{\Lambda\setminus\Lambda'}(\sigma)$$

be the Boltzmann–Gibbs state which interpolates between the system with Gaussian disorder and the system with a flipped disorder in the region Λ' ($H^{(1)}_{\Lambda'}$ and $H^{(2)}_{\Lambda'}$ are two independent copies of the Hamiltonian in the subvolume Λ', $H_{\Lambda\setminus\Lambda'}(\sigma)$ is the Hamiltonian in the remaining part of the volume; they are all independent). Then, the following identities hold

$$\lim_{\Lambda,\Lambda'\nearrow\mathbb{Z}^d}\int_0^{\pi}\!\!\int_0^{\pi} dt\,ds\ \sin^2(s-t)\left[\left\langle(c^{\Lambda'}_{1,2})^2\right\rangle_{t,s} - 2\left\langle c^{\Lambda'}_{1,2}c^{\Lambda'}_{2,3}\right\rangle_{s,t,s} + \left\langle c^{\Lambda'}_{1,2}c^{\Lambda'}_{3,4}\right\rangle_{t,s,s,t}\right] = 0 \tag{5.182}$$

where $\langle(c^{\Lambda'}_{1,2})^2\rangle_{t,s}$ (and analogously for the other terms) is the overlap of region $\Lambda' \subseteq \Lambda$ in the quenched state constructed from the interpolating Boltzmann–Gibbs state, e.g.

$$\langle(c^{\Lambda'}_{1,2})^2\rangle_{t,s} = \mathbb{E}\left[\omega_t\omega_s\left(c^2_{\Lambda'}(\sigma,\tau)\right)\right].$$

Proof The proof is obtained from a suitable combination of results from the previous sections. For a parameter $t \in [0, \pi]$, we consider the interpolating random pressure

$$\mathcal{P}(t) = \ln \sum_{\sigma \in \Sigma_\Lambda} e^{x_\sigma(t) + H_{\Lambda \setminus \Lambda'}(\sigma)} \tag{5.183}$$

where

$$x_\sigma(t) = \cos(t) H^{(1)}_{\Lambda'}(\sigma) + \sin(t) H^{(2)}_{\Lambda'}(\sigma)$$

with $H^{(1)}_{\Lambda'}(\sigma)$ and $H^{(2)}_{\Lambda'}(\sigma)$ being two independent copies of the Hamiltonian for the subvolume $\Lambda' \subseteq \Lambda$. The boundary value of the interpolating pressure gives the random pressure of the original system when $t = 0$, and the random pressure of the system with the couplings J flipped on the subvolume Λ' when $t = \pi$, i.e.

$$\mathcal{P}(0) = \mathcal{P}_\Lambda,$$
$$\mathcal{P}(\pi) = \mathcal{P}_{\Lambda,\Lambda'}.$$

Application of Lemma 5.38 with $\xi_\sigma = H^{(1)}_{\Lambda'}(\sigma)$ and $\eta_\sigma = H^{(2)}_{\Lambda'}(\sigma)$ gives

$$\begin{aligned} &\mathrm{Var}(\mathcal{P}_\Lambda - \mathcal{P}_{\Lambda,\Lambda'}) \\ &= \Lambda' \int_0^\pi \int_0^\pi dt\, ds\, \cos(s - t) \langle c^{\Lambda'}_{1,2} \rangle_{t,s} + (\Lambda')^2 \int_0^\pi \int_0^\pi dt\, ds\, \sin^2(s - t) \\ &\times \left[\left\langle (c^{\Lambda'}_{1,2})^2 \right\rangle_{t,s} - 2 \left\langle c^{\Lambda'}_{1,2} c^{\Lambda'}_{2,3} \right\rangle_{s,t,s} + \left\langle c^{\Lambda'}_{1,2} c^{\Lambda'}_{3,4} \right\rangle_{t,s,s,t} \right] \end{aligned} \tag{5.184}$$

(the presence of the additional term $H_{\Lambda \setminus \Lambda'}(\sigma)$ in the random interpolating pressure does not change the result in the lemma, as long as the quenched state is correctly interpreted). On the other hand, Lemma 5.39 tells us that $\mathrm{Var}(\mathcal{P}_\Lambda - \mathcal{P}_{\Lambda,\Lambda'})$ is bounded above by a constant times the subvolume Λ'. As a consequence, the statement of the theorem follows. □

Remark 5.41 When expressed in terms of the spin variables, the polynomial in the integral (5.182) involves generalized truncated correlation functions. Indeed we have the following expressions

$$\omega_{t,s}\left((C^{\Lambda'}_{1,2})^2\right) = \sum_{X,Y \subset \Lambda'} \Delta^2_X \Delta^2_Y \omega_t\left(\sigma^{(1)}_X \sigma^{(1)}_Y\right) \omega_s\left(\sigma^{(2)}_X \sigma^{(2)}_Y\right)$$

$$\omega_{s,t,s}\left(C^{\Lambda'}_{1,2} C^{\Lambda'}_{2,3}\right) = \sum_{X,Y \subset \Lambda'} \Delta^2_X \Delta^2_Y \omega_s\left(\sigma^{(1)}_X\right) \omega_t\left(\sigma^{(2)}_X \sigma^{(2)}_Y\right) \omega_s\left(\sigma^{(3)}_Y\right)$$

$$\omega_{t,s,s,t}\left(C^{\Lambda'}_{1,2} C^{\Lambda'}_{3,4}\right) = \sum_{X,Y \subset \Lambda'} \Delta^2_X \Delta^2_Y \omega_t\left(\sigma^{(1)}_X\right) \omega_s\left(\sigma^{(2)}_X\right) \omega_s\left(\sigma^{(3)}_Y\right) \omega_t\left(\sigma^{(4)}_Y\right)$$

thus

$$\omega_{t,s}\left(\left(c_{1,2}^{\Lambda'}\right)^2\right) - 2\,\omega_{s,t,s}\left(c_{1,2}^{\Lambda'}c_{2,3}^{\Lambda'}\right) + \omega_{t,s,s,t}\left(c_{1,2}^{\Lambda'}c_{3,4}^{\Lambda'}\right)$$
$$= \frac{1}{|\Lambda'|^2}\sum_{X,Y\subset\Lambda'}\Delta_X^2\Delta_Y^2\,[\omega_t(\sigma_X\sigma_Y)-\omega_t(\sigma_X)\omega_t(\sigma_Y)]\,[\omega_s(\sigma_X\sigma_Y)-\omega_s(\sigma_X)\omega_s(\sigma_Y)] \tag{5.185}$$

where replica indexes have been dropped.

Example 5.42 For the Edwards–Anderson model, which is obtained with $\Delta_X^2 = 1$ if $X \in B' = \{(n,n') \in \Lambda' \times \Lambda', |n-n'| = 1\}$, and $\Delta_X^2 = 0$ otherwise, the linear combination (5.185) of the moments of the link overlap in the region Λ' is written in terms of truncated correlation functions, that is

$$\omega_{t,s}\left(\left(c_{1,2}^{\Lambda'}\right)^2\right) - 2\,\omega_{s,t,s}\left(c_{1,2}^{\Lambda'}c_{2,3}^{\Lambda'}\right) + \omega_{t,s,s,t}\left(c_{1,2}^{\Lambda'}c_{3,4}^{\Lambda'}\right)$$
$$= \frac{1}{|\Lambda'|^2}\sum_{b,b'\in B'}[\omega_t(\sigma_b\sigma_{b'})-\omega_t(\sigma_b)\omega_t(\sigma_{b'})]\,[\omega_s(\sigma_b\sigma_{b'})-\omega_s(\sigma_b)\omega_s(\sigma_{b'})]\,, \tag{5.186}$$

with $\sigma_b = \sigma_n\sigma_n'$, if $b = (n,n') \in B'$. See also Temesvari (2007).

Example 5.43 (Triviality of the random field model) We compute here explicitly the expression appearing in Theorem 5.40

$$\left\langle c_{1,2}^2\right\rangle_{t,s} - 2\langle c_{1,2}c_{2,3}\rangle_{s,t,s} + \langle c_{1,2}c_{3,4}\rangle_{t,s,s,t} \tag{5.187}$$

in the simple case of a non-interacting random field. We will show that this linear combination of overlap moments vanishes pointwise for all values of t and s. We will then deduce the triviality of the order parameter for the random field model.

We consider two families J_i and $\tilde{J}_i$ for $i = 1,\ldots,N$ of independent normally distributed centered random variables with variance 1:

$$\mathbb{E}\left[J_iJ_j\right] = \mathbb{E}\left[\tilde{J}_i\tilde{J}_j\right] = \delta_{i,j}, \quad \mathbb{E}\left[J_i\tilde{J}_j\right] = 0, \tag{5.188}$$

and the random field Hamiltonians

$$\xi_\sigma = \sum_{i=1}^N J_i\sigma_i, \qquad \eta_\sigma = \sum_{i=1}^N \tilde{J}_i\sigma_i. \tag{5.189}$$

where $\sigma_i = \pm 1$. We have that $\xi = \{\xi_\sigma\}_\sigma$ and $\eta = \{\eta_\sigma\}_\sigma$ are two independent centered Gaussian families (each having $n = 2^N$ elements indexed by configurations σ, where N is the volume) with covariance structures given by:

$$\begin{aligned}\mathbb{E}\left[\xi_\sigma \xi_\tau\right] &\equiv \mathcal{C}_{\sigma,\tau} = Nq(\sigma,\tau),\\ \mathbb{E}\left[\eta_\sigma \eta_\tau\right] &\equiv \mathcal{C}_{\sigma,\tau} = Nq(\sigma,\tau),\\ \mathbb{E}\left[\xi_\sigma \eta_\tau\right] &= 0,\end{aligned} \tag{5.190}$$

where $q(\sigma,\tau)$ is the *site overlap* of the two configurations σ and τ:

$$q(\sigma,\tau) = \frac{1}{N}\sum_{i,j=1}^{N} \sigma_i \tau_j. \tag{5.191}$$

The interpolating Hamiltonian:

$$x_\sigma(t) = \cos(t)\xi_\sigma + \sin(t)\eta_\sigma, \tag{5.192}$$

which is a stationary Gaussian process with the same distribution of ξ and η:

$$\mathbb{E}\left[x_\sigma(t) x_\tau(t)\right] = Nq(\sigma,\tau), \tag{5.193}$$

defines the quenched deformed state on the replicated system, whose averages are denoted with the usual notation, e.g. $\langle - \rangle_{t,s}$, $\langle - \rangle_{s,t,s} \dots$ Considering the random field spin glass with Hamiltonian (5.189), in the limit $N \to \infty$ and for all values of t and s:

$$\gamma_1 \langle q_{1,2}^2 \rangle_{t,s} + \gamma_2 \langle q_{1,2} \rangle_{t,s}^2 + \gamma_3 \langle q_{1,2} q_{2,3} \rangle_{s,t,s} + \gamma_4 \langle q_{1,2} q_{3,4} \rangle_{t,s,s,t} = 0 \tag{5.194}$$

for any choice of real $\gamma_1, \gamma_2, \gamma_3, \gamma_4$ with $\gamma_1 + \gamma_2 + \gamma_3 + \gamma_4 = 0$.

This can be seen by a simple explicit computation on the free random field model (see Contucci *et al.* (2009a)):

$$\langle C_{1,2} \rangle_{t,s}^2 = \sum_{i=1}^{N} \left(\mathbb{E}\left[\tanh(G_i(t)) \tanh(G_i(s))\right]\right)^2 + \mathcal{Q}_N(t,s), \tag{5.195}$$

$$\left\langle C_{1,2}^2 \right\rangle_{t,s} = 1 + \mathcal{Q}_N(t,s), \tag{5.196}$$

$$\langle C_{1,2} C_{2,3} \rangle_{s,t,s} = \sum_{i=1}^{N} \mathbb{E}\left[\tanh^2(G_i(s))\right] + \mathcal{Q}_N(t,s), \tag{5.197}$$

$$\langle C_{12} C_{34} \rangle_{t,s,s,t} = \sum_{j=1}^{N} \mathbb{E}\left[\tanh^2(G_j(t)) \tanh^2(G_j(s))\right] + \mathcal{Q}_N(t,s), \tag{5.198}$$

where

$$G_i(t) = \cos(t)J_i + \sin(t)\tilde{J}_i, \tag{5.199}$$

and $\mathcal{Q}_N(t,s)$ is a term of order N^2. Thus:

$$\gamma_1\langle C_{1,2}^2\rangle_{t,s} + \gamma_2\langle C_{1,2}\rangle_{t,s}^2 + \gamma_3\langle C_{1,2}C_{2,3}\rangle_{s,t,s} + \gamma_3\langle C_{1,2}C_{3,4}\rangle_{t,s,s,t}$$

$$= \gamma_1 + \gamma_2\sum_{i=1}^{N}(\mathbb{E}\left[\tanh(G_i(t))\tanh(G_i(s))\right])^2 + \gamma_3\sum_{i=1}^{N}\mathbb{E}\left[\tanh^2(G_i(s))\right]$$

$$+\gamma_4\sum_{j=1}^{N}\mathbb{E}\left[\tanh^2(G_j(t))\tanh^2(G_j(s))\right] + (\gamma_1+\gamma_2+\gamma_3+\gamma_4)\mathcal{Q}_N(t,s),$$

i.e. the linear combination of the covariance matrix moments is of order N. Thus, since $|\tanh(x)| < 1$, we have

$$\left|\gamma_1\langle C_{1,2}^2\rangle_{t,s} + \gamma_2\langle C_{1,2}\rangle_{t,s}^2 + \gamma_3\langle C_{1,2}C_{2,3}\rangle_{s,t,s} + \gamma_4\langle C_{1,2}C_{3,4}\rangle_{t,s,s,t}\right|$$
$$\leqslant |\gamma_1| + (|\gamma_2|+|\gamma_3|+|\gamma_4|)N,$$

which can be rewritten, using the overlaps $q_{1,2},\ q_{2,3},\ q_{3,4}$ between replicas, as

$$\left|\gamma_1\langle q_{1,2}^2\rangle_{t,s} + \gamma_2\langle q_{1,2}\rangle_{t,s}^2 + \gamma_3\langle q_{1,2}q_{2,3}\rangle_{s,t,s} + \gamma_4\langle q_{1,2}q_{3,4}\rangle_{t,s,s,t}\right|$$
$$\leqslant \frac{|\gamma_2|+|\gamma_3|+|\gamma_4|}{N} + \frac{|\gamma_1|}{N^2}. \tag{5.200}$$

Among the relations in equations (5.194), in the thermodynamic limit, we find the identity of Theorem 5.40 for the values $\gamma_1 = 1, \gamma_2 = 0, \gamma_3 = -2, \gamma_4 = 1$:

$$\langle q_{1,2}^2\rangle_{t,s} - 2\langle q_{1,2}q_{2,3}\rangle_{s,t,s} + \langle q_{1,2}q_{3,4}\rangle_{t,s,s,t} = 0 \tag{5.201}$$

and the Ghirlanda–Guerra identities: for $\gamma_1 = 1,\ \gamma_2 = 1,\ \gamma_3 = -2,\ \gamma_4 = 0$ we find

$$\langle q_{1,2}q_{2,3}\rangle_{s,t,s} = \frac{1}{2}\langle q_{1,2}^2\rangle_{t,s} + \frac{1}{2}\langle q_{1,2}\rangle_{t,s}^2; \tag{5.202}$$

for $\gamma_1 = 1, \gamma_2 = 2, \gamma_3 = 0, \gamma_4 = -3$ we find

$$\langle q_{1,2}q_{3,4}\rangle_{s,t,s} = \frac{1}{3}\langle q_{1,2}^2\rangle_{t,s} + \frac{2}{3}\langle q_{1,2}\rangle_{t,s}^2. \tag{5.203}$$

Using (5.202) and (5.203) we can express (5.201) as:

$$\langle q_{1,2}^2\rangle_{t,s} - 2\langle q_{1,2}q_{2,3}\rangle_{s,t,s} + \langle q_{1,2}q_{3,4}\rangle_{t,s,s,t} = \frac{1}{3}\left(\langle q_{1,2}^2\rangle_{t,s} - \langle q_{1,2}\rangle_{t,s}^2\right). \tag{5.204}$$

The identity derived from the flip of the coupling thus implies a trivial order parameter distribution. Indeed, since the identity (5.201) is true for every t and s, we can choose $t = s = 0$ and then the interpolating states reduce to the usual quenched Boltzmann–Gibbs state. From Eq. (5.204), we deduce a trivial overlap distribution.

6
Numerical simulations

Abstract

In this chapter we present some results on numerical simulation in three-dimensional spin glasses. In fact, there are problems for which the analytical approach is out of reach. In such cases numerical simulations are a source of hints and suggestions. Their robustness is based on the fact that the asymptotic properties of the model for large volumes are identified. After describing the standard algorithm used (parallel tempering) we address the following problems: overlap equivalence among site and link overlaps, ultrametricity or hierarchical organization of the equilibrium states, decay of correlations, pure state identification, energy interfaces, and stiffness exponents related to the lower critical dimension.

6.1 Introduction

Numerical simulations have played a crucial role in the development of spin glass theory, especially in those cases in which exact results are unavailable due to formidable mathematical difficulties. Some of the first systematic works were by Bray and Moore (1984), Ogielski and Morgenstern (1985), and Bhatt and Young (1988). For a recent review, see Marinari *et al.* (1997) and the references therein. It is also interesting to check Newman and Stein (1996) and Marinari *et al.* (2000) for examples of the difficulties of reconciling theory with numerical simulations in finite-dimensional spin glasses.

The general outcome of these (and other) numerical studies is that many features of the mean-field theory are seen in the ever larger finite-volume systems accessible to numerical simulations. However, no definite conclusions can be reached by means of the sole use of computers. A statement describing (metaphorically) a similar situation comes from the Heraclitus fragment 93 (see Plutarch, On the Pythian Oracle, 404E in Kirk *et al.* (1995)) about the Delphi Oracle who *neither conceals nor reveals the truth, but only hints at it.*

Numerical simulations of complex systems like spin glasses are extremely delicate and hard to perform, as well as to interpret, but they provide useful benchmarks for the theory. On the other hand, the richness of the involved physical picture has, in turn, been the source of inspiration for extremely powerful new algorithms that have fertilized other fields of interest – such as computational science and optimization problems. Examples of those algorithms range from the *simulated annealing* (Kirkpatrick *et al.* (1983)) to the *parallel tempering* (Marinari and Parisi (1992); Hukushima and Nemoto (1996)), and up to the most recents contributions of *belief propagation* and *survey propagation* (Mézard and Parisi (2001); Mézard *et al.* (2002); Mézard and Montanari (2009)).

Going back to the problem of the nature of the low temperature phase of a realistic model of a spin glass, one must say that the problem is completely open, as described in the commentary Bouchaud (2005). In this chapter we will review some selected results and investigate a few issues related to the nature of the spin glass phase in the three-dimensional Edwards–Anderson model (overlap equivalence, ultrametricity, decay of correlations, energy interfaces). The indication that emerges from these studies is that on the limited sizes accessible to simulations, mean-field features are clearly seen. The main question that these studies raise is: which of these features remain in the thermodynamic limit? The answer to this question could also come from the study of dynamical properties of spin glasses.

6.2 Simulations with real replicas

The standard tool used nowadays to simulate spin glasses is the *parallel tempering* (PT) algorithm, introduced in Marinari and Parisi (1992); Hukushima and Nemoto (1996). This is a Markov chain Monte Carlo method which is especially designed for the simulation of statistical mechanics systems with an equilibrium measure with many different configurations having relevant weights. In this section, we briefly review the method and its practical implementation in a spin system.

For a given disorder realization, i.e. assigned couplings J, we would like to sample spin configurations from the product Boltzmann–Gibbs equilibrium distribution $\mu_{\Lambda,\beta_1,\ldots,\beta_R}$ of $R \in \mathbb{N}$ copies of a spin system in a volume Λ, each copy having its own temperature β_j with $j = 1, \ldots, R$. We denote the spin configurations of the replicated system by $\boldsymbol{\sigma} = (\sigma^{(1)}, \ldots, \sigma^{(R)})$ with $\sigma^{(j)} \in \Sigma_\Lambda = \{-1, +1\}^{|\Lambda|}$. The probability of a configuration is then

$$\mu_{\Lambda,\beta_1,\ldots,\beta_R}(\boldsymbol{\sigma}) = \prod_{j=1}^{R} \frac{e^{-\beta_j H_\Lambda(\sigma^{(j)})}}{Z_\Lambda(\beta_j)}. \tag{6.1}$$

The sampling from this measure will be obtained by simulating a Markov chain that has (6.1) as an invariant measure. It is convenient to formulate the dynamics in continuous time via the generator

$$Lf(\boldsymbol{\sigma}) = \sum_{\boldsymbol{\eta}} r(\boldsymbol{\sigma}, \boldsymbol{\eta})(f(\boldsymbol{\eta}) - f(\boldsymbol{\sigma})) \tag{6.2}$$

where $f : \Sigma_\Lambda^R \to \mathbb{R}$ is a generic function and $r(\boldsymbol{\sigma}, \boldsymbol{\eta})$ represents the rate of transition from the configuration $\boldsymbol{\sigma}$ to the configuration $\boldsymbol{\eta}$. Those rates are chosen so that $\mu_{\Lambda,\beta_1,\ldots,\beta_R}$ satisfies detailed balance, i.e.

$$\mu_{\Lambda,\beta_1,\ldots,\beta_R}(\boldsymbol{\sigma})\, r(\boldsymbol{\sigma}, \boldsymbol{\eta}) = \mu_{\Lambda,\beta_1,\ldots,\beta_R}(\boldsymbol{\eta})\, r(\boldsymbol{\eta}, \boldsymbol{\sigma}) \tag{6.3}$$

therefore implying stationarity of the probability measure (6.1). In the PT algorithm, two types of transitions are allowed:

- *standard Metropolis*

$$\boldsymbol{\sigma} = (\sigma^{(1)}, \ldots, \sigma^{(j)}, \ldots, \sigma^{(R)}) \to \boldsymbol{\eta} = (\sigma^{(1)}, \ldots, \sigma^{(j),i}, \ldots, \sigma^{(R)}) \tag{6.4}$$

  where the configuration $\sigma^{(j)}$ of the jth copy is changed to the configuration $\sigma^{(j),i}$ with a flipped spin at position $i \in \Lambda$. This happens at rate

$$r(\boldsymbol{\sigma}, \boldsymbol{\eta}) = \min\{1, e^{-\beta_j \Delta H}\} \tag{6.5}$$

  with

$$\Delta H = H_\Lambda(\sigma^{(j),i}) - H_\Lambda(\sigma^{(j)}); \tag{6.6}$$

- *replica exchange*

$$\boldsymbol{\sigma} = (\sigma^{(1)}, \ldots, \sigma^{(j)}, \sigma^{(j+1)}, \ldots, \sigma^{(R)}) \to \boldsymbol{\eta} = (\sigma^{(1)}, \ldots, \sigma^{(j+1)}, \sigma^{(j)}, \ldots, \sigma^{(R)}) \tag{6.7}$$

  where the spin configurations $\sigma^{(j)}$ of the jth copy and $\sigma^{(j+1)}$ of the adjacent $(j+1)$th copy are exchanged. This happens at rate

$$r(\boldsymbol{\sigma}, \boldsymbol{\eta}) = \min\{1, e^{-\Delta}\} \tag{6.8}$$

  with

$$\Delta = (\beta_{j+1} - \beta_j)(H_\Lambda(\sigma^{(j)}) - H_\Lambda(\sigma^{(j+1)})). \tag{6.9}$$

It is easy to check that the rates (6.5) and (6.8) both satisfy detailed balance condition (6.3). More general formulations of the PT algorithm consider general spin flips in Metropolis-type transitions (i.e. flipping not just one spin, but an entire set of spins) or exchanging two arbitrary copies (which are not necessarily adjacent) in the replica exchange transitions.

Denoting by $\{\sigma(t)\}_{t\geqslant 0}$ the continuous-time Markov chain defined by the generator (6.2) whose rates are those of the PT algorithm, the expectation of the single-copy observable $\mathcal{O} : \Sigma_\Lambda \to \mathbb{R}$ with respect to the random Boltzmann–Gibbs measure at inverse temperature β_j defined by

$$\omega_{\beta_j}(\mathcal{O}) = \sum_{\sigma} \mathcal{O}(\sigma) \frac{e^{-\beta_j H_\Lambda(\sigma)}}{Z_\Lambda(\beta_j)}, \tag{6.10}$$

is estimated by the time average

$$\bar{\mathcal{O}}_{\beta_j}(t) = \frac{1}{t} \int_0^t dt' \, \mathcal{O}(\sigma^{(j)}(t')). \tag{6.11}$$

By the ergodic theorem for irreducible aperiodic Markov chains, one has (almost surely with respect to $\mu_{\Lambda,\beta_1,\dots,\beta_R}$)

$$\lim_{t\to\infty} \bar{\mathcal{O}}_{\beta_j}(t) = \omega_{\beta_j}(\mathcal{O}). \tag{6.12}$$

Analogous results apply to multiple-copy observables.

Remark 6.1 In the implementation of the PT algorithm on a computer, one usually considers discrete time dynamics. A standard way to do so is to alternate a complete Metropolis *sweep* of the lattice (i.e. sequential or random flip of all the spins of the lattice independently for each copy) with a complete *swap* of the copies (i.e. sequential or random exchange of all adjacent copies).

Remark 6.2 The study of rate of convergence of the distribution $\nu^{(t)}$ at time t to the stationary distribution $\mu_{\Lambda,\beta_1,\dots,\beta_R}$ is a classical problem of probability theory. For irreducible aperiodic continuous time Markov chains on a finite state space, the total variational distance between the two distributions is controlled by an inequality such as

$$||\nu^{(t)} - \mu_{\Lambda,\beta_1,\dots,\beta_R}|| \leqslant M\rho^t \tag{6.13}$$

for some $0 < \rho < 1$ and $M < \infty$. The constant ρ is related to the spectral properties of the generator. However, such an inequality is not always useful to establish the convergence speed, because the computation of the spectral gap might be very difficult for high-dimensional configuration spaces.

In practical cases (i.e. running a simulation) it is notoriously difficult to decide whether or not the Markov chain is close to its equilibrium. To improve the chances of reaching the stationarity, an initial "thermalization time" is usually added at the beginning of a simulation. Namely, the equilibrium average $\omega_{\beta_j}(\mathcal{O})$ is better estimated by

$$\bar{\mathcal{O}}_{\beta_j}(t_0, t) = \frac{1}{t} \int_{t_0}^{t_0+t} dt' \, \mathcal{O}(\sigma^{(j)}(t')). \tag{6.14}$$

Specific equilibration criteria will be illustrated in the following sections.

6.3 Overlap equivalence

In this section we analyze two important quantities for the spin glass phase of the three-dimensional Gaussian Edwards–Anderson model with no external field. One is the *link overlap* which is the Hamiltonian covariance and, with its probability distribution, it encodes the properties of the system. The other is the *standard overlap* which is related to the Edwards–Anderson order parameter. Physicists have different views of how the two quantities should behave when the large-volume limit is reached. In particular, the so called trivial–nontrivial (TNT) picture predicts that one variable (the standard overlap) is nontrivial while the other (the link overlap) is trivial.

We present the results from the paper Contucci *et al.* (2006) where, by use of numerical simulations, the fluctuations of the two random variables were studied. First we analyze the correlation coefficient and find that the two quantities are uncorrelated above the critical temperature. Below the critical temperature we find that the link overlap has vanishing fluctuations for fixed values of the square standard overlap and large volumes. The results show that the conditional variance scales to zero in the thermodynamic limit. This implies that, if one of the two random variables tends to a trivial one (i.e. delta-like distributed), so does the other one and, by consequence, the TNT picture should be dismissed. The functional relation between the two variables – which turns out to be a monotonically increasing function – is identified using the method of least squares. The results show that the two overlaps are completely equivalent in the description of the low temperature phase of the Edwards–Anderson model.

To detect the nature of the low temperature phase of short-range spin glasses, Edwards and Anderson (1975) originally proposed an order parameter: the disorder average of the local squared magnetization

$$\mathbb{E}\left[\omega^2(\sigma_i)\right] = \mathbb{E}\left[\left[\frac{\sum_\sigma \sigma_i e^{-\beta H_\sigma}}{\sum_\sigma e^{-\beta H_\sigma}}\right]^2\right], \tag{6.15}$$

which coincides with the quenched expectation of the local *standard* overlap of two spin configurations, drawn according to two copies of the equilibrium state carrying identical disorder

$$\langle q_i \rangle = \mathbb{E}\left[\frac{\sum_{\sigma,\tau} \sigma_i \tau_i \, e^{-\beta(H_\sigma + H_\tau)}}{\sum_{\sigma,\tau} e^{-\beta(H_\sigma + H_\tau)}} \right]. \tag{6.16}$$

The previous parameter should reveal the presence of frozen spins in random directions at low temperatures. While that choice of local observable is quite natural, it is far from being unique; one can consider, for instance, the two-point function $\mathbb{E}\left[\omega^2(\sigma_i \sigma_j)\right] = \langle q_i q_j \rangle$. In the case of nearest-neighbor overlap correlation function, this yields to the quenched average of the local *link overlap*.

When summed over the whole volume, link and standard overlaps give rise to a priori different global order parameters. In the mean-field case, the two have a very simple relationship: in the Sherrington–Kirkpatrick (SK) model, for instance, it turns out that the link overlap coincides with the square power of the standard overlap. But in general, especially in the finite-dimensional case of nearest-neighbor interaction like the Edwards–Anderson (EA) model, the two previous quantities have a different behavior with respect to spin flips: when summed over regions, the first undergoes changes of volume sizes after spin flips, while the second is affected only by surface terms.

As it was explained in Chapter 1, their role is especially different from the mathematical point of view. The square power of the standard overlap represents, in fact, the covariance of the Hamiltonian function for the SK model, while the link overlap represents the covariance for the EA model. The two different overlap definitions are naturally related to two different notions of distance among spin configurations. It would be interesting to establish if the two distances are equivalent for the equilibrium measure in the large-volume limit and, if so, to what extent (see Parisi and Ricci-Tersenghi (2000) for a broad discussion on *overlap equivalence* and its relation with ultrametricity).

From the geometrical point of view, the two distances could in fact be simply equivalent in preserving neighborhoods (topological equivalence) or they could preserve order among distances (metric equivalence). Clearly the metric equivalence implies a topological one, but not vice versa. Topological equivalence would be violated if the functional relation result between the two distances turns out be discontinuous, while metric equivalence would be violated by a non-monotonic function. The outcomes of this section show numerical evidence for a metric equivalence.

It is important to mention that the different properties of the two overlaps have been discussed in relation to all the proposed spin glass pictures: droplet (Fisher

Table 6.1 *Parameters of the simulations: system size (L), number of sweeps used for thermalization (N_{therm}), number of sweeps for measurement of the observables (N_{equil}), number of disorder realizations (N_{real}), number of temperature values allowed in the PT procedure (n_β), temperature increment (δT), minimum (T_{min}) and maximum (T_{max}) temperature values.*

L	N_{therm}	N_{equil}	N_{real}	n_β	δT	T_{min}	T_{max}
3–6	50 000	50 000	2 048	19	0.1	0.5	2.3
8	50 000	50 000	2 680	19	0.1	0.5	2.3
10	70 000	70 000	2 048	37	0.05	0.5	2.3
12	70 000	70 000	2 048	37	0.05	0.5	2.3

and Huse (1986)), mean-field (Mézard *et al.* (1987)), and TNT (Krzakala and Martin (2000); Palassini and Young (2000)). The distributions of the two overlaps are expected, respectively, to be delta-like (trivial distribution, droplet theory), to have support on a finite interval (nontrivial distribution, mean-field theory), or to have different behavior depending on which overlap is considered (trivial link overlap distribution, nontrivial standard overlap distribution, TNT theory). For a discussion of the relation between overlaps distribution and excitations/interfaces, see also Newman and Stein (2001, 2003a).

In the work Contucci *et al.* (2006), the EA model in three dimensions was considered, with Gaussian couplings and zero external magnetic field in periodic boundary conditions. The relative fluctuations of the link overlap with respect to the square of standard overlap were studied. The PT algorithm was used to investigate lattices with linear sizes from $L = 3$ to $L = 12$. For each size, at least 2048 disorder realizations were simulated. For the larger sizes, 37 temperature values in the range $0.5 \leqslant T \leqslant 2.3$ were used. The choice of the lowest temperature was related to the possibility of thermalizing the large sizes, but the results were perfectly compatible with those obtained by Marinari and Parisi (2001) at temperature $T = 0$ (see below). The thermalization in the PT procedure was tested by checking the symmetry of the probability distribution for the standard overlap q under the transformation $q \to -q$. Moreover, for the Gaussian coupling case, another thermalization test is available: the internal energy can be calculated both as the temporal mean of the Hamiltonian and, by exploiting integration by parts, as an expectation of a simple function of the link overlap. It was checked that both measurements converged to the same value with the above thermalization steps. All the parameters used in the simulations are reported in Table 6.1.

For a three-dimensional cube Λ of volume $N = L^3$, considering the standard overlap q (Eq. (1.7)) and the link overlap Q (Eq. (1.4)), we first investigate the behavior of the correlation coefficient between q^2 and Q

$$\rho = \frac{\langle (q^2 - \langle q^2 \rangle)(Q - \langle Q \rangle) \rangle}{\sqrt{\langle (q^2 - \langle q^2 \rangle)^2 \rangle \langle (Q - \langle Q \rangle)^2 \rangle}}. \tag{6.17}$$

This quantity will tell us in which range of temperatures the two random variables are correlated. In that range we can then further investigate the nature of the mutual correlation by studying their joint distribution and, in particular, the conditional distribution $P(Q|q^2)$ of Q at fixed values of q^2. One is interested in discovering if a functional relation exists between the two quantities, i.e. if the variance of the conditional distribution shrinks to zero at large volumes, and around which curve the conditional distribution peaks. The conditional probability mass function is given by:

$$P(Q|q^2) = \frac{P(Q, q^2)}{P(q^2)} = \frac{\mathbb{E}\left[\frac{\sum_{\sigma,\tau} \delta(Q - Q_{\sigma,\tau})\delta(q^2 - q^2_{\sigma,\tau})e^{-\beta[H_\sigma + H_\tau]}}{\sum_{\sigma,\tau} e^{-\beta[H_\sigma + H_\tau]}}\right]}{\mathbb{E}\left[\frac{\sum_{\sigma,\tau} \delta(q^2 - q^2_{\sigma,\tau})e^{-\beta[H_\sigma + H_\tau]}}{\sum_{\sigma,\tau} e^{-\beta[H_\sigma + H_\tau]}}\right]}. \tag{6.18}$$

For this conditional distribution, one could compute the generic kth moment

$$G_k(q^2) := \int_{-1}^{1} Q^k P(Q|q^2) dQ. \tag{6.19}$$

Important quantities are the conditional mean

$$G(q^2) := G_1(q^2) \tag{6.20}$$

and the conditional variance

$$\text{Var}(Q|q^2) := G_2(q^2) - G_1^2(q^2). \tag{6.21}$$

The method of least squares immediately entails that the mean $G(q^2)$ is the best estimator for the functional dependence of Q in terms of q^2. In fact, given any function $h(q^2)$, the mean of $(Q - h(q^2))^2$ according to the joint distribution $P(Q, q^2)$, is

$$\sum_{i,j} \left(Q_i - h(q_j^2)\right)^2 P(Q_i, q_j^2) = \sum_j P(q_j^2) \sum_i \left(Q_i - h(q_j^2)\right)^2 P(Q_i|q_j^2),$$

where the sums cover all possible values of the random vector (Q, q^2), which are finite and many on the simulated finite system. Therefore, to minimize the mean it suffices to minimize the inner sum, i.e. to choose $h(q^2)$ as the mean $G(q^2)$ of Q with respect to the conditional probability mass function (6.18).

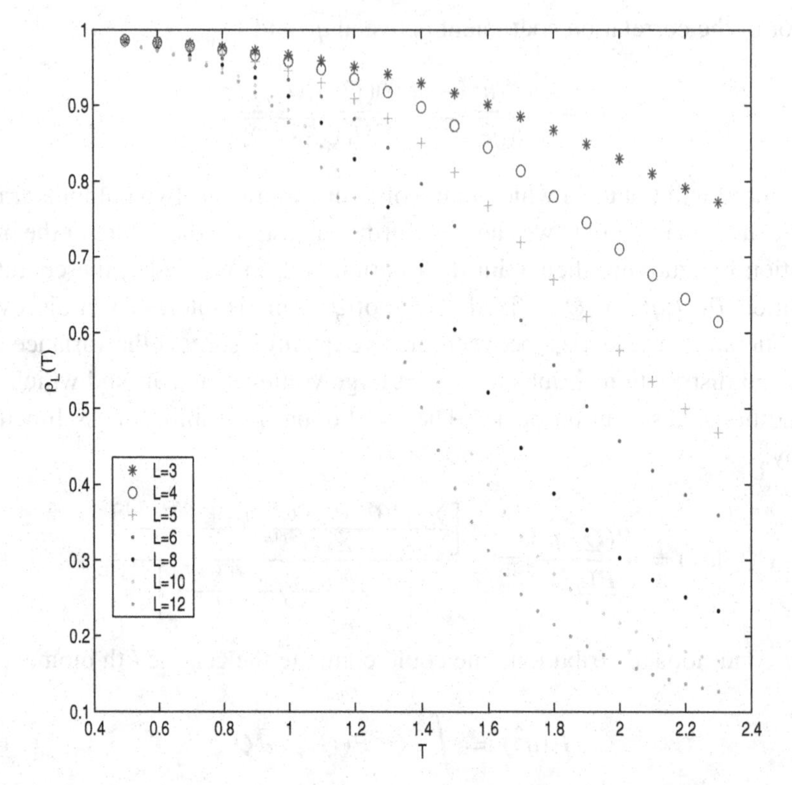

Figure 6.1 Correlation $\rho_L(T)$ as a function of the temperature T for different sizes L of the system.

The scaling properties of the conditional variance (6.21) and the functional dependence (6.20) provide important information about the low temperature phase of the model. Indeed, a vanishing variance in the thermodynamic limit implies that the two random variables Q and q^2 do not fluctuate with respect to each other. If the functional dependence $G(q^2)$ among the two is a one-to-one increasing function, then it follows that the marginal probability distributions for the standard and link overlaps must have similar properties. In particular, if one of the two is supported over a point, so must the other.

Let us now describe the numerical results. Figure 6.1 shows the correlation between the square standard overlap and the link overlap. Equation (6.17) is plotted for different sizes of the system as a function of the temperature. It is clear from the figure that as the system size increases, the correlation remains strong in the low temperature region, but becomes weaker in the high temperature region. A sudden change in the infinite-volume behavior of ρ can be expected to occur close to the critical temperature T_c of the model. The best estimate available in the literature – obtained through the analysis of Binder parameter curves of the

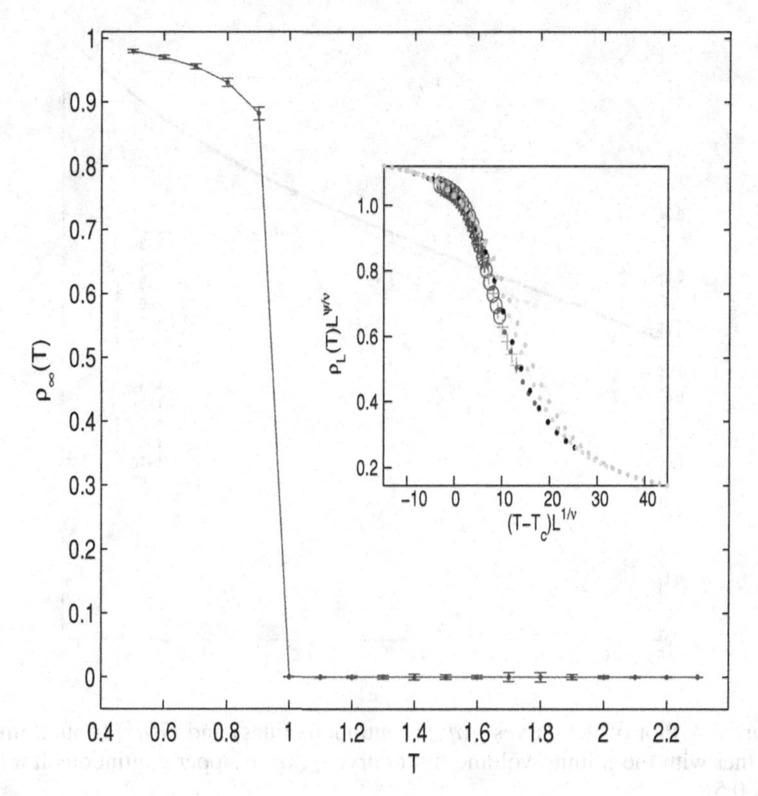

Figure 6.2 $\rho_\infty(T)$ as a function of the temperature T. Inset shows the finite size scaling.

variable q^2 for different system sizes – gives $T_c = 0.95 \pm 0.04$ (Marinari *et al.* (1997)). Independent data reproduces an estimate $T_c = 0.95 \pm 0.03$.

For each temperature, a fit of the data was done for ρ to the infinite-volume limit. Different scalings for the data were tried, both exponential $\rho_L(T) = \rho_\infty(T) + a(T)e^{b(T)L}$ and power law $\rho_L(T) = \rho_\infty(T) + \alpha(T)L^{\beta(T)}$, where T denotes the temperature. The interesting information is contained in the asymptotic value $\rho_\infty(T)$. The normalized χ^2 was measured for different values of $\rho_\infty(T)$ in the range $[0, \min_L \rho_L(T)]$, keeping $a(T)$ and $b(T)$ (or $\alpha(T)$ and $\beta(T)$) as free parameters. In the region $T \geq 1.0$, it was found that χ^2 attains its minimal value for $\rho_\infty(T) = 0$. For $T \leqslant 0.9$, χ^2 develops a sharp minimum corresponding to values $\rho_\infty(T) \neq 0$. The whole plot of the curve $\rho_\infty(T)$, obtained from the best fit, is represented in Figure 6.2. The inset of the same figure shows the standard finite-size scaling of the data. In the plot of $\rho_L(T)L^{\psi/\nu}$ versus the scaling variable $(T - T_c)L^{1/\nu}$, a good scaling is obtained using $T_c = 0.95$, $\nu = 0.71$, $\psi = 0.038$. The discrepancy between $\nu = 0.71$ and the value $\nu = 2.0$ of Marinari *et al.* (1997) might to be attributed to the nonlinear relation between Q and q^2 (see below). Figure 6.2 tells

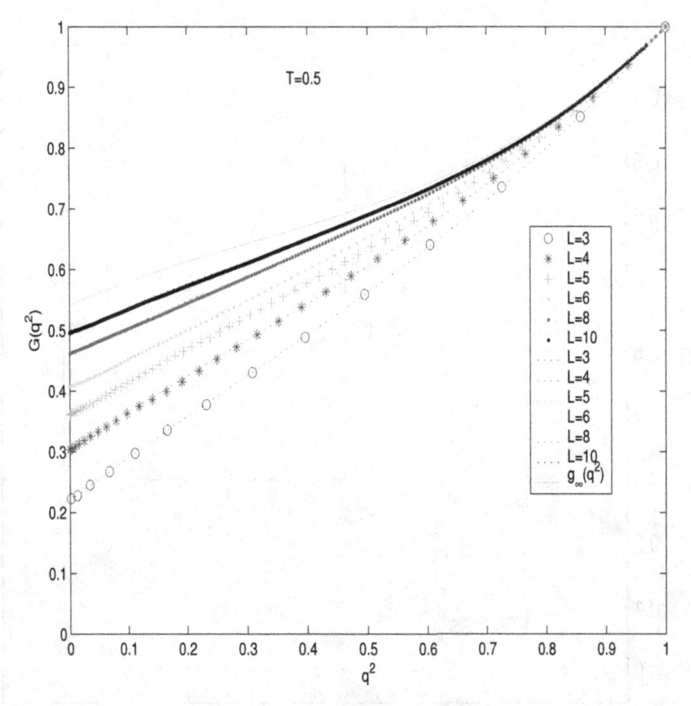

Figure 6.3 Plot of the curves $g(q^2)$ (continuous lines) and $G(q^2)$ (dotted lines), together with the infinite-volume limit curve $g_\infty(q^2)$ (upper continuous line) for $T = 0.5$.

us that in the high-temperature phase, the two random variables (standard and link overlap) are asymptotically uncorrelated, while in the low temperature phase they display a non-vanishing correlation: within the discrete set of temperature values we have available, the temperature at which the correlation coefficient starts to be different from zero is in close agreement with the estimated critical value of the model.

It was considered then the problem of studying the functional dependence (if any) between the two random variables Q and q^2 in the low temperature region. The points in Figure 6.3 show the function $G(q^2)$ of Eq. (6.20) for different system sizes at $T = 0.5$, well below the critical temperature. A third order approximation of the form

$$Q = g(q^2) = a + bq^2 + cq^4 + dq^6 \tag{6.22}$$

was also studied. Since one must have $Q = 1$ for $q^2 = 1$, this actually implies $d = 1 - a - b - c$. The coefficients $a_{L,T}, b_{L,T}, c_{L,T}$ have been obtained by the least square method and then fitted to the infinite-volume limit. The results are

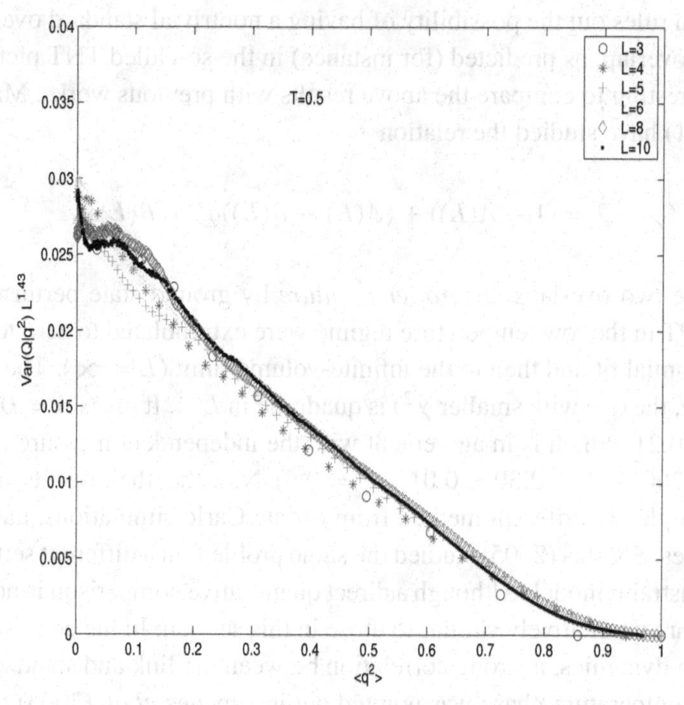

Figure 6.4 $L^{1.43}\mathrm{Var}_L(Q|q^2)$ as a function of q^2 for the temperature $T = 0.5$ and for different sizes L.

shown as continuous lines in Figure 6.3. The good superposition of the curves to the data for $G(q^2)$ indicates that the functional dependence between the two overlaps is already well approximated at the third order.

Finally, the variance defined in Eq. (6.21) was measured at low temperatures. One may see that the distribution concentrates around its mean value for large volumes. The trend toward a vanishing variance for infinite system sizes is very clear. All available temperatures below T_c were analyzed and it was found that the best fit of $\mathrm{Var}_L(Q|q^2)$, in terms of χ^2, was obtained by a power law of the form $a(T)L^{-b(T)} + c$ which gave $c = 0$ for every value of the temperature. For the lowest available temperature $T = 0.5$, shown in Figure 6.4 where the data for $\mathrm{Var}_L(Q|q^2)L^{1.43}$ are plotted for different system sizes L, all the different curves collapse to a single one. The data for other temperature values behave similarly, the only difference being that the coefficient $b(T)$ increases with the temperature (it stays in the range $[1.43, 1.74]$ for $T \in [0.5, 0.9]$). This result has quite strong consequences because it says that the two random variables Q and q^2 cannot have different triviality properties: if one of them is trivial (delta-like distributed) the scaling law for their conditional variance implies that the other is also trivial. Such a result then rules out the possibility of having a nontrivial standard overlap with a trivial link overlap, as predicted (for instance) in the so-called TNT picture.

It is interesting to compare the above results with previous works. Marinari and Parisi (2001) have studied the relation

$$Q = (1 - A(L)) + (A(L) - B(L))q^2 + B(L)q^4 \tag{6.23}$$

between the two overlaps *at zero temperature* by ground state perturbation. The data from PT in the low temperature regime were extrapolated to zero temperature by a polynomial fit and then to the infinite-volume limit ($L = \infty$). The best fit for $L = \infty$ (i.e. the one with smaller χ^2) is quadratic in L^{-1}. It gives $A = 0.30 \pm 0.05$ with $\chi^2 = 0.21$, which is in agreement with the independent measure of Marinari and Parisi (2001) ($A = 0.30 \pm 0.01,\ \chi^2 = 0.6$). Note that their results are obtained using a completely different method from Monte Carlo simulations, namely exact ground states. Sourlas (2005) studied the same problem in a different setting, called the soft constraint model. Although a direct quantitative comparison is not possible, his results are qualitatively similar to those in this section. In the context of out-of-equilibrium dynamics, a strong correlation between the link and standard overlaps in the low temperature phase was pointed out in Jimenez *et al.* (2003).

In conclusion, the result of Contucci *et al.* (2006) shows quite clearly that, within the tested system sizes, the two order parameters – namely the square of the standard overlap, and the link overlap, are equivalent as far as the quenched equilibrium state is concerned. In view of our result, the proposed pictures which assign different behavior to the two overlap distributions, in particular the TNT picture, should be rejected.

6.4 Ultrametricity

Ultrametricity is a crucial ingredient in the replica computations of the SK model (cf. Section 1.7), as well as a guiding principle for the rigorous proof of its free energy density formula (cf. Sections 4.12 and 4.13). While the rigorous proof of ultrametricty is still in progress for mean-field models (see, in particular, the paper Panchenko (2011a)), it is a widely accepted property from the theoretical physics perspective. Its relevance in finite-dimensional systems is nonetheless still an open matter, and the subject of intense investigations and debates in the theoretical and mathematical physics communities.

The popular view of ultrametricity can be stated as a very striking property for physical systems: essentially, it says that the equilibrium configurations of a large system can be classified in a taxonomic (hierarchical) way: configurations are grouped in states, states are grouped in families, families are grouped in super-families. Such an equilibrium ultrametricity has a correspondence in the existence of widely separated timescales in the dynamics, typical of a glassy system.

Following the work of Contucci *et al.* (2007), we present here numerical simulations that support the ultrametric property in three-dimensional spin glasses. Part of the difficulty of such a study is because ultrametricity should be, at best, exact when the volume of the system goes to infinity, and therefore simulations done on a limited range of volumes are difficult to interpret. In Contucci *et al.*, systems ranging from 4^3 to 20^3 were studied, extending the volume range used in previous simulations by around an order of magnitude.

In Euclidean spaces, sampling three points uniformly and independently, one almost surely gets scalene triangles. Ultrametricity implies instead that sampling three configurations independently – with respect to their common Boltzmann–Gibbs state and averaging over the disorder – the distribution of the distances between them is supported, in the limit of very large systems, only on equilateral and isosceles triangles (with no contribution from scalene triangles). In a generic situation, the relative weight of equilateral and isosceles triangles is arbitrary. However, it is well established in stochastically stable systems that a quarter of the triangles are equilateral, while three-quarters are isosceles.

The property of ultrametricity and the nontrivial structure of the overlap distribution are the characterizing features of the mean-field picture and are mutually intertwined: a trivial (delta-like) overlap probability distribution, like the one predicted in the droplet theory, is not compatible in fact with the previous ultrametric structure because it predicts only equilateral triangles.

In Contucci *et al.* (2007), the Edwards–Anderson model in the three-dimensional cubic lattice with $\pm J$ random interactions was studied (a numerical analysis in four dimensions was done in Cacciuto *et al.* (1997)). Very strong indications were found in favor of ultrametricity which turned out to be reached at large volumes with exactly the form predicted by the mean-field theory. Consequently, a robust indication against droplet theory was founded. (For a study of dynamical ultrametricity, and for the relation between statics and dynamics in spin glasses, see Franz and Ricci-Tersenghi (2000); Franz *et al.* (1998).) Violations of ultrametricity in finite volumes were shown to have a clear tendency to vanish as the system size increased. The analytical predictions of the ultrametric replica symmetry breaking ansatz were verified to be correct up to the tested sizes. Those results contradicted previous findings – for instance, those of Hed *et al.* (2004) for much smaller volumes (up to 8^3), in which lack of ultrametricity was claimed.

The distribution of the overlaps were investigated with a multi-spin coding and a PT algorithm (the parameters used in the simulations are reported in Table 6.2). Thermalization was checked by verifying that the result would have been the same

Table 6.2 *Parameters of the simulations: system size (L), number of sweeps used for thermalization (N_{therm}), number of disorder realizations (N_{real}), number of temperature values allowed in the PT procedure (n_β), minimum ($T_{\min}$) and maximum ($T_{\max}$) temperature values.*

L	N_{therm}	N_{real}	n_β	$T_{\min}$	$T_{\max}$
4	1 047 552	1 280	25	0.7	2.1
6	1 047 552	1 280	25	0.7	2.1
8	1 047 552	1 280	25	0.7	2.1
10	1 047 552	1 280	25	0.7	2.1
12	1 047 552	896	25	0.7	2.1
16	2 096 128	1 216	25	0.7	2.1
18	2 096 128	768	49	0.7	2.1
20	4 193 280	512	103	0.7	2.1

(inside the error bar) by taking simulations a factor of 4 shorter. According to the literature, the system has a transition temperature T_c of 1.15. The lowest temperature used was 0.7, i.e. about 0.6 T_c.

The triple of random variables $(c_{1,2}, c_{2,3}, c_{3,1})$, with $0 \leqslant c_{i,j} \leqslant 1$, representing the overlaps among three copies of the system, is called *stochastically stable and ultrametric* when, defining $\chi(c) = \int_0^c P(c')dc'$, where $P(c)$ is the probability distribution of c, the joint probability density function has the following structure:

$$
\begin{aligned}
P_3(c_{1,2}, c_{2,3}, c_{3,1}) = {} & \frac{1}{2} P(c_{1,2})\chi(c_{1,2})\delta(c_{1,2} - c_{2,3})\delta(c_{2,3} - c_{3,1}) \\
& + \frac{1}{2} P(c_{1,2})P(c_{2,3})\theta(c_{1,2} - c_{2,3})\delta(c_{2,3} - c_{3,1}) \\
& + \frac{1}{2} P(c_{2,3})P(c_{3,1})\theta(c_{2,3} - c_{3,1})\delta(c_{3,1} - c_{1,2}) \\
& + \frac{1}{2} P(c_{3,1})P(c_{1,2})\theta(c_{3,1} - c_{1,2})\delta(c_{1,2} - c_{2,3}). \qquad (6.24)
\end{aligned}
$$

Thinking of the quantities cs as 1 minus the sides of a triangle, Eq. (6.24) says that only equilateral (first term on the righthand side) and isosceles (last three terms) triangles are allowed; the scalene triangles have zero probability. Equation (6.24) implies that the distribution of the three random variables $u = \min(c_{1,2}, c_{2,3}, c_{3,1})$, $v = \text{med}(c_{1,2}, c_{2,3}, c_{3,1})$ and $z = \max(c_{1,2}, c_{2,3}, c_{3,1})$ is

$$
\rho(u, v, z) = \frac{1}{2}\chi(u)P(u)\delta(v - u)\delta(z - v) + \frac{3}{2}P(z)P(v)\theta(z - v)\delta(v - u), \qquad (6.25)
$$

and from this, one deduces that the distribution of the two differences $x = v - u$, $y = z - v$ is

$$\tilde{\rho}(x, y) = \delta(x)\left[\frac{1}{4}\delta(y) + \frac{3}{2}\theta(y)\int_y^1 P(a)P(a-y)da\right], \qquad (6.26)$$

whose marginals are

$$\tilde{\rho}(x) = \delta(x), \qquad (6.27)$$

and

$$\tilde{\rho}(y) = \frac{1}{4}\delta(y) + \frac{3}{2}\theta(y)\int_y^1 P(a)P(a-y)da. \qquad (6.28)$$

For every function of two spin configurations $\mathcal{O}(\sigma, \tau)$, the physical model induces a probability distribution by the formula

$$\mathcal{P}_3(O_{1,2}, O_{2,3}, O_{3,1}) = \langle \delta(O_{1,2} - \mathcal{O}(\sigma, \tau))\delta(O_{2,3} - \mathcal{O}(\tau, \gamma))\delta(O_{3,1} - \mathcal{O}(\gamma, \sigma))\rangle, \qquad (6.29)$$

where σ, τ, γ denote three different equilibrium configurations. As usual, the brackets $\langle \cdot \rangle$ denote the average over the disorder of the thermal average over the Boltzmann–Gibbs distribution. An observable $\mathcal{O}$ is said to be ultrametric if its joint quenched distribution over three copies (6.29) is given, in the limit of large volumes, by formula (6.24).

We examine here the ultrametricity property as it was studied for the standard overlap q (Eq. (1.7)) and the link overlap Q (Eq. (1.4)) of the Edwards–Anderson model on a three-dimensional cube Λ of volume $N = L^3$. Very strong evidence was found that for large volumes, the link overlap has the ultrametric structure of Eq. (6.24) (in a zero magnetic field, the system is invariant under a global sign change of all the spins). The same results were also found for the standard overlap, the only difference being that it has a symmetric distribution in the interval $[-1, 1]$, and the triangle distribution is built on it by suitable contributions of the positive and negative values; see formula (6.30) below. The results for the link overlap are illustrated first, followed by those for the standard overlap.

For the link overlap, the structure of the distribution for the two random variables $X = Q_{\text{med}} - Q_{\text{min}}$ and $Y = Q_{\text{max}} - Q_{\text{med}}$, (the Qs represent the largest, medium, and smallest values of the link overlap among three copies of the system) was investigated. The numerical data were compared to the formulas (6.27) and (6.28).

The variances of the two variables were found to have a totally different behavior. The left panel of Figure 6.5 contains the plot of $\text{Var}(X)/\text{Var}(Q)$ and the right

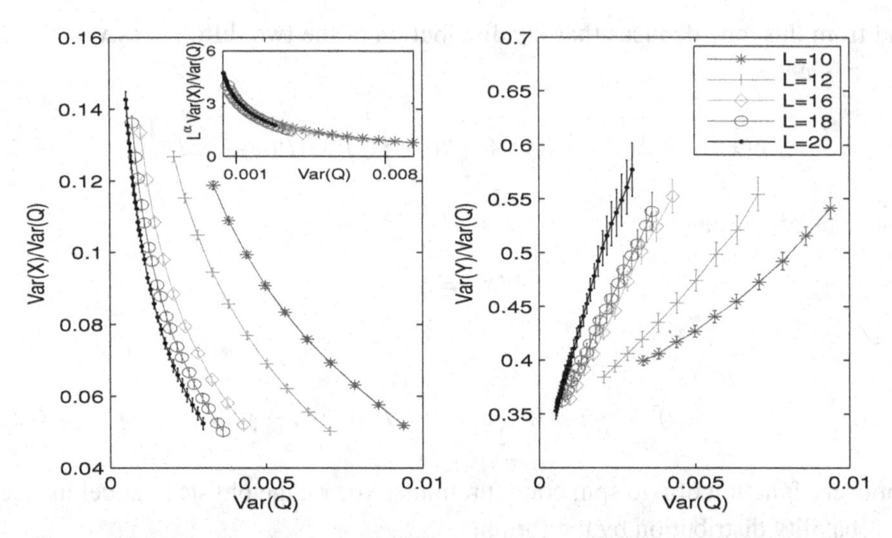

Figure 6.5 Normalized variances of the two random variables $X = Q_{\text{med}} - Q_{\text{min}}$ (left) and $Y = Q_{\text{max}} - Q_{\text{med}}$ (right) as a function of $\text{Var}(Q)$. The inset (left panel) shows that the scaling law for $\alpha = 1.18$, i.e. $L^{\alpha}\text{Var}(X)/\text{Var}(Q)$ is L-independent.

panel of $\text{Var}(Y)/\text{Var}(Q)$, both as functions of $\text{Var}(Q)$. This parametrization is more convenient with respect to the usual one using temperature, because it allows the extraction of more information on size dependence through scaling laws. This is because for $\text{Var}(X)$,$\text{Var}(Y)$, and $\text{Var}(Q)$, size dependence changes with the temperature T. In particular, within the temperature range that we have taken into account, the quantity $\text{Var}(Q)$ decreases monotonically with the temperature. Figure 6.5 clearly shows that while the variance of X is shrinking to zero, the variance of Y is growing with the volume. Moreover, the variance of X satisfies a scaling law with very good accuracy: $\text{Var}(X)/\text{Var}(Q)$ scales like $L^{-1.18}$ (see inset) while there is no scaling law for the Y variable.

Figure 6.6 displays the data histograms for two system sizes of $L = 12$ and $L = 20$ for X (dotted line) and Y (line comprising of circles) variables at temperature $T = 0.7$. They show that $\mathcal{P}(X)$, the empirical distribution of X, is much more concentrated close to zero, while $\mathcal{P}(Y)$ is spread on a larger scale. The function $\tilde{\rho}(Y)$ provides a test of consistency with formula (6.28). The $\tilde{\rho}(Y)$ plots have been obtained using the data histograms of X to represent the delta function (6.27), and the experimental data for the distribution of Q inside the convolution. The two curves superimpose each other closely. This is a self-consistent test of the weights one-quarter and three-quarters.

The analysis for the standard overlap must be done in a different way, since in a zero magnetic field, the $\mathbb{Z}_2$ symmetry is not broken. Given the three standard overlaps $q_{1,2}$, $q_{2,3}$, $q_{3,1}$, their probability measure is *a priori* supported on $[-1, 1]^3$.

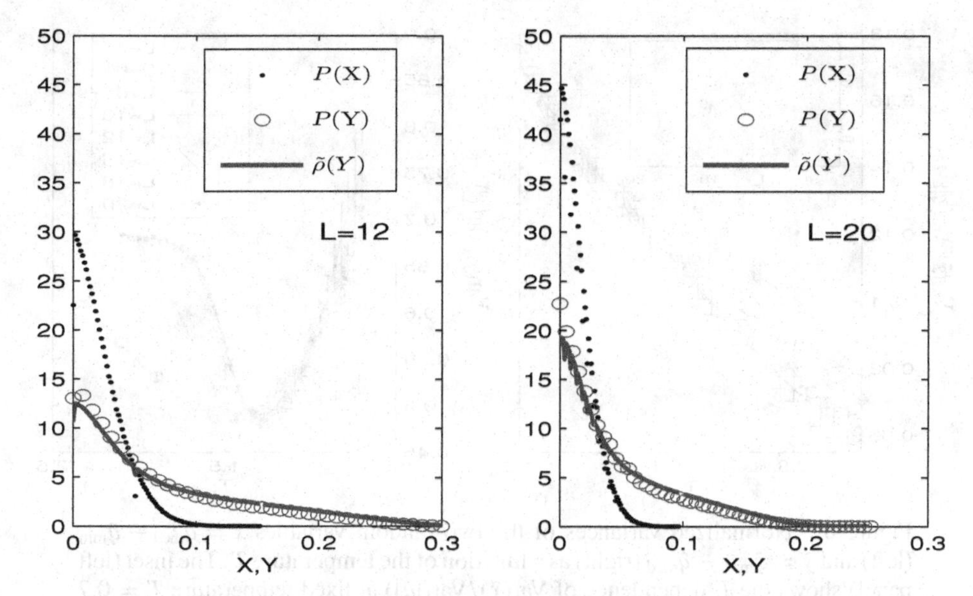

Figure 6.6 Empirical distributions $\mathcal{P}(X)$ and $\mathcal{P}(Y)$ for $X = Q_{\text{med}} - Q_{\text{min}}$ and $Y = Q_{\text{max}} - Q_{\text{med}}$ for the two system sizes ($L = 12$ and $L = 20$) at temperature $T = 0.7$. $\tilde{\rho}(Y)$ shows the distribution of Y computed from formula (6.28) using experimental data for $\mathcal{P}(Q)$ and approximating the delta function with the histogram of X.

Reflection invariance ($q_{i,j} \to \alpha_i q_{i,j} \alpha_j$, with $\alpha = \pm 1$) implies that the probability measure is a sum of two orbits, one for $S = \text{sign}(q_{1,2}\, q_{2,3}\, q_{3,1}) > 0$ and the other for $S < 0$. The mean-field theory predicts that only non-frustrated triples ($S > 0$) contribute to the triangle distribution, namely:

$$\begin{aligned}
\bar{P}_3(q_{1,2}, q_{2,3}, q_{3,1}) = \frac{1}{4} \big[& P_3(q_{1,2}, q_{2,3}, q_{3,1})\theta(q_{1,2})\theta(q_{2,3})\theta(q_{3,1}) \\
& + P_3(-q_{1,2}, -q_{2,3}, q_{3,1})\theta(-q_{1,2})\theta(-q_{2,3})\theta(q_{3,1}) \\
& + P_3(q_{1,2}, -q_{2,3}, -q_{3,1})\theta(q_{1,2})\theta(-q_{2,3})\theta(-q_{3,1}) \\
& + P_3(-q_{1,2}, q_{2,3}, -q_{3,1})\theta(-q_{1,2})\theta(q_{2,3})\theta(-q_{3,1}) \big] . \qquad (6.30)
\end{aligned}$$

To check the validity of the previous formula, it is convenient to introduce the new random variables

$$\tilde{q}_{max} = \max(|q_{1,2}|, |q_{2,3}|, |q_{3,1}|) \qquad (6.31)$$

$$\tilde{q}_{med} = \text{med}(|q_{1,2}|, |q_{2,3}|, |q_{3,1}|) \qquad (6.32)$$

$$\tilde{q}_{min} = \text{sign}(q_{1,2}q_{2,3}q_{3,1}) \min(|q_{1,2}|, |q_{2,3}|, |q_{3,1}|) \qquad (6.33)$$

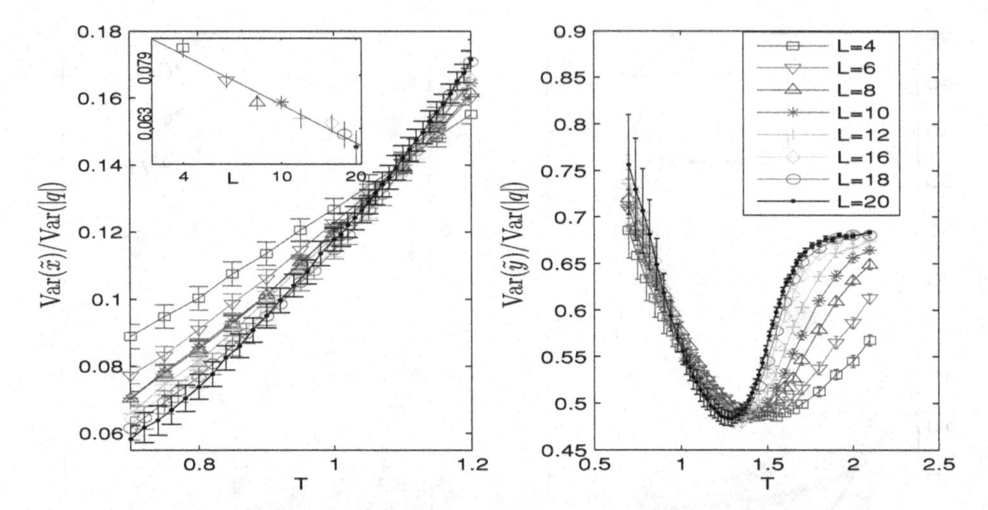

Figure 6.7 Normalized variances of the two random variables $\tilde{x} = \tilde{q}_{\rm med} - \tilde{q}_{\rm min}$ (left) and $\tilde{y} = \tilde{q}_{\rm max} - \tilde{q}_{\rm med}$ (right) as a function of the temperature T. The inset (left panel) shows the L-dependence of $\mathrm{Var}(\tilde{x})/\mathrm{Var}(|q|)$ at fixed temperature $T = 0.7$ on a log–log scale, together with the best fit: $\mathrm{Var}(\tilde{x})/\mathrm{Var}(|q|) \sim aL^b$, $a = 0.12$, $b = -0.23$.

and verify that their distribution is given by (6.24). The numerical results are illustrated in Figure 6.7. The left panel shows how the normalized variance of the variable $\tilde{x} = \tilde{q}_{\rm med} - \tilde{q}_{\rm min}$ has a clear tendency to vanish for temperatures below the critical point. The inset displays the log–log plot of $\mathrm{Var}(\tilde{x})/\mathrm{Var}(|q|)$ as a function of L at the lowest available temperature $T = 0.7$. At the critical point, this quantity is size-invariant instead, as predicted by mean-field theory. A totally different behavior is found for the variable $\tilde{y} = \tilde{q}_{\rm max} - \tilde{q}_{\rm med}$, where below the critical temperature the normalized variance is increasing, but is still size-invariant at criticality.

The contribution of the frustrated triples was explicitly investigated by plotting the quantity

$$S^{(-)} = \frac{\int_{-1}^{0} d\tilde{q}_{\rm min}\, p(\tilde{q}_{\rm min})\, \tilde{q}_{\rm min}^2}{\int_{-1}^{1} d\tilde{q}_{\rm min}\, p(\tilde{q}_{\rm min})\, \tilde{q}_{\rm min}^2}, \tag{6.34}$$

where $p(\tilde{q}_{\rm min})$ is the probability distribution of $\tilde{q}_{\rm min}$. The left panel of Figure 6.8 clearly shows that the distribution of $\tilde{q}_{\rm min}$ is almost completely supported on the positive interval, and that the negative values are concentrated near zero (for similar quantities and other three-replica observables, see Iniguez *et al.* (1996)). This implies that the contribution associated with the frustrated orbit ($S < 0$) is very small at large volumes.

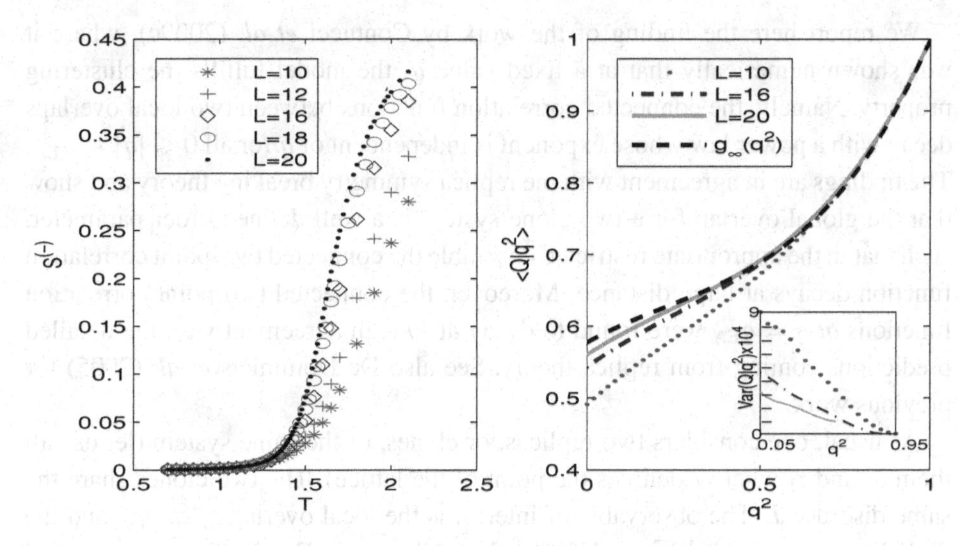

Figure 6.8 Left panel: the average value of $S^{(-)}$ (defined in the text) as a function of T. Right panel: conditional expectation $G(q^2) = \langle Q|q^2\rangle$ and conditional variance $\mathrm{Var}(Q|q^2)$ (inset) of the random variable Q, given q^2, where Q is the link overlap and q^2 is the square of the standard overlap, for different system sizes at temperature $T = 0.7$.

The equivalent behavior of link and standard overlaps is indeed expected because of the findings of Section 6.3, where it was shown that link and standard overlaps are mutually non-fluctuating for the case of Gaussian couplings. The right panel of Figure 6.8 shows, for the model with $\pm J$ investigated within this section, the analysis of the relative fluctuation and functional dependence of the two overlaps. It is shown that the function $G(q^2)$ of Eq. (6.20), i.e. the expected value of the link overlap for an assigned value of the standard overlap, for different system sizes at $T = 0.7$, with a fit to the infinite-volume limit $g_\infty(q^2)$. The conditional variance of Q, given q^2, Eq. (6.21), is displayed in the inset and shows a trend toward a vanishing value for infinite system sizes.

6.5 Decay of correlations

In the mean-field theory of spin glasses there are intensive quantities that also fluctuate in the thermodynamic limit. This is the effect of the coexistence of many equilibrium states. The correlation functions inside a given state are expected to have a power law behavior: below the critical temperature, spin glasses are always in a critical state (many glassy systems share this behavior). The analysis performed in Section 6.3 for the bond spin variables at distance one (i.e. nearest neighbor) in the conditioned state with a fixed value of the site overlap, can be performed for arbitrary distances.

We report here the finding of the work by Contucci *et al.* (2009c), where it was shown numerically that at a fixed value q, the model fulfills the clustering property. Namely, the connected correlation functions between two local overlaps decay with a power law whose exponent is independent of q for all $0 \leqslant |q| < q_{\text{EA}}$. The findings are in agreement with the replica symmetry breaking theory and show that the global overlap for a two clone system is a well defined order parameter, such that in the appropriate restricted ensemble the connected two-point correlation function decays at large distance. Moreover, the connected two-point correlation functions *at* $q = q_{\text{EA}}$ were found to decay at $1/x$, in agreement with the detailed predictions coming from replica theory. See also De Dominicis *et al.* (2005) for previous work.

As usual, one considers two replicas, or clones, of the same system (let us call them σ_i and τ_i where i denotes the point of the lattice). The two clones share the same disorder J. The observable of interest is the local overlap $q_i \equiv \sigma_i \tau_i$ and the global overlap $q \equiv V^{-1} \sum_i q_i$, V being the total volume. For the three-dimensional Edwards–Anderson model at zero magnetic field, all simulations confirm that the probability distribution $P_J(q)$ is nontrivial in the thermodynamic limit – it changes from system to system and its average over the disorder, that we denote by $P(q) = \mathbb{E}[P_J(q)]$, is nontrivial. It has support in the region from $-q_{\text{EA}}$ to q_{EA}, q_{EA}, being the overlap of two generic configurations belonging to the same state. It is usually assumed that the function $P(q)$ has, in the infinite-volume limit, a delta function singularity at $q = q_{\text{EA}}$, that appears as a peak in finite-volume systems. In the presence of multiple states, the most straightforward approach consists of identifying the clustering states (i.e. those where the connected correlation functions go to zero at a large distance) and introducing an order parameter that identifies the different states. This task is extremely difficult in a random system where the structure of the states depends on the instance of the system. However, the replica theory is able to make predictions without finding out explicitly the set of states. At this point the introduction of the two clones plays a crucial role. Indeed, if the global overlap q has a preassigned value, the theory predicts that the correlations of local overlaps q_i go to zero at large distances. In other words, q is a good order parameter.

We define $\omega_q^J(O)$ to be the expectation value of the observable O in the J-dependent Gibbs ensemble, restricted to those configurations of the two clones that have global overlap fixed to the value q. We define the average expectations values $\langle\langle O \rangle\rangle_q$ as the weighted average over the systems of restricted expectation values

$$\langle\langle O \rangle\rangle_q = \frac{\mathbb{E}\left[P_J(q)\omega_q^J(O)\right]}{\mathbb{E}\left[P_J(q)\right]}. \tag{6.35}$$

Alternatively, one could define the expectations values $\langle O \rangle_q$ as the unweighted average over the systems of restricted expectation values:

$$\langle O \rangle_q = \mathbb{E}\left[\omega_q^J(O)\right]. \tag{6.36}$$

It is convenient to follow definition (6.35) because it is very easy to implement numerically: all the configurations produced in a numerical simulation can be classified according to their overlaps, independent of the system they come from, and the average of each class can easily be performed.

The main statement of the replica symmetry breaking theory is that the q-dependent connected correlation functions go to zero when computed in the ensemble $\langle\langle \cdot \rangle\rangle_q$, i.e. the states $\langle\langle \cdot \rangle\rangle_q$ are clustering. By the use of Jensen's inequality, one can show that if clustering property holds in the conditioned state $\langle\langle \cdot \rangle\rangle_q$ defined by (6.35), then it also holds in the conditioned state $\langle \cdot \rangle_q$ defined by (6.36).

The procedure is very similar to the one used in ferromagnetic models to construct clustering states (for example in the Ising case, by considering averages with positive or negative total magnetization). The overlap constrained state is not clustering when the equilibrium state is locally unique (apart from a global change of signs). Indeed, mean-field spin glasses are the only known example of a system where the clustering states are labeled by a continuously changing order parameter in the absence of a continuous symmetry, such as rotations or translations.

It is convenient to recall the known theoretical results for the connected correlation functions in the case of short-range Ising spin glasses. Taking the two points in the lattice $(x, 0, 0)$ and $(0, 0, 0)$ we introduce

$$G(x|q) = \langle \sigma_x \tau_x \sigma_0 \tau_0 \rangle_q, \quad C(x|q) = G(x|q) - q^2 \tag{6.37}$$

and their Fourier transforms $\tilde{C}(k|q)$. The simplest predictions are obtained starting from mean-field theory and computing the first nontrivial term (De Dominicis *et al.* (1998)). Neglecting logarithms one has in the small k region

$$\begin{aligned}
\tilde{C}(k|q) &\propto k^{-4} \quad \text{for } q = 0, \\
\tilde{C}(k|q) &\propto k^{-3} \quad \text{for } 0 < q < q_{\mathrm{EA}}, \\
\tilde{C}(k|q) &\propto k^{-2} \quad \text{for } q = q_{\mathrm{EA}}, \\
\tilde{C}(k|q) &\propto (k^2 + \xi(q)^{-2})^{-1} \quad \text{for } q > q_{\mathrm{EA}}.
\end{aligned} \tag{6.38}$$

These results are supposed to be exact at large distances in sufficiently high dimensions, i.e. for $d > 6$. One may be surprised to find a result for $q > q_{\mathrm{EA}}$ because the function $P(q)$ is zero in this region in the infinite-volume limit. However for finite systems $P(q)$ is different from zero for any q, although it is very small in the region $q > q_{\mathrm{EA}}$ (Franz *et al.* (1992); Billoire *et al.* (2003)). For $q > q_{\mathrm{EA}}$ an analytical computation of the function $\tilde{C}(k|q)$ has not yet been done; however, it is reasonable to assume that the leading singularity near to $k = 0$ in the complex plane is a single pole, leading to an exponentially decaying correlation function. For example, an exponential decrease of the correlation is present in the Heisenberg model if one constrains the modulus of magnetization to have a value greater than the equilibrium value. On the contrary, if the modulus of the magnetization is less than the equilibrium value, the connected correlation function does not go to zero at large distances.

When the dimensions become less than six, one can rely on the perturbative expansion in $\epsilon = 6 - d$, where only the first order is (partially) known (see Temesvari and De Dominicis (2002); De Dominicis and Giardina (2006)). It seems that predictions at $q = q_{\text{EA}}$ should not change and the form of the $k = 0$ singularity at $q = q_{\text{EA}}$ (i.e. when the two clones belong to the same state) remains k^{-2}, as for Goldstone bosons. On the contrary, the $k = 0$ singularities at $q < q_{\text{EA}}$ should change. One can expect that

$$\tilde{C}(k|q) \propto k^{-\tilde{\alpha}(q)} \quad \text{for} \quad 0 \leqslant q < q_{\text{EA}}. \tag{6.39}$$

These perturbative results are the only information one has on the form of $\tilde{\alpha}(q)$. The simplest scenario would be that $\tilde{\alpha}(q)$ is discontinuous at $q = 0$ and constant in the region $0 < q < q_{\text{EA}}$. It would be fair to say that there is no strong theoretical evidence for the constancy of $\tilde{\alpha}(q)$ in the region $0 < q < q_{\text{EA}}$, apart from generic universality arguments. On the contrary, the discontinuity at $q = 0$ could persist in dimensions not too smaller than six, and disappear at lower dimensions, as supported by the data presented below in $d = 3$.

In the three-dimensional case in configuration space, one should have:

$$C(x|q) \propto x^{-\alpha(q)} \quad \text{for} \quad 0 \leqslant q \leqslant q_{\text{EA}} \tag{6.40}$$

with $\alpha(q_{\text{EA}}) = 1$ (indeed, in general, one has that $\alpha(q) = D - \tilde{\alpha}(q)$). For $q > q_{\text{EA}}$ the correlation should go to zero faster than a power: one tentatively assumes that

$$C(x|q) \propto x^{-1} \exp(-x/\xi(q)) \quad \text{for} \quad q_{\text{EA}} < q, \tag{6.41}$$

– other behaviors are of course possible.

In Contucci *et al.* (2009c), the properties of the two overlaps' connected correlation functions in the three-dimensional Edwards–Anderson model with $J_{i,j} = \pm 1$ (symmetrically distributed) were studied. Cubic lattice systems of side L for $L = 4, 6, 8, 10, 12, 16, 20$ were analyzed. Periodic boundary conditions were used and the simulation parameters were the same as in Table 6.2. The results below hold at temperature $T = 0.7$, while the critical temperature is about $T_c = 1.11$. The configurations created during the numerical simulations were classified according

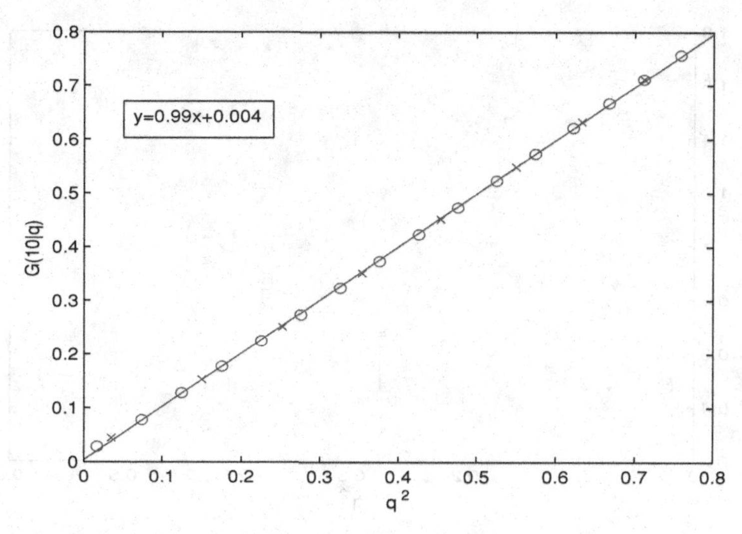

Figure 6.9 The correlation function $G(10|q)$ (at distance 10), averaged in 20 bins (round points) and ten bins (crosses) of q^2, versus the average of q^2 inside the bin. The data are for a system of size 20 and the straight line is the best fit to the data.

to the value of the global overlap q. Since the properties of the configurations are invariant under the symmetry $(q \to -q)$, the configurations were classified into 20 equidistant bins in q^2, e.g. the first bin contained all the configurations where $0 < q^2 < 1/20$. In this way the correlations $C(x|q)$ were computed. The correlations were measured only along the axes of the lattice: x is an integer restricted to the range $[0, L/2]$. As a control, the same operation was done with 10 bins, obtaining similar results.

Firstly it was verified that the connected correlations vanish for large systems. In Figure 6.9 the correlation $G(10|q)$ versus the average of q^2 in the bin is plotted for $L = 20$. One see that the two quantities coincide. The data show strong evidence for the vanishing of the connected two-point correlation function. The predictions of replica theory are $G(10|q) = q^2$, neglecting small corrections going to zero with the volume.

Further information can be extracted from the data. The analysis of the data should be done in a different way in the two regions ($0 \leqslant q^2 \leqslant q_{\mathrm{EA}}^2$ and $q_{\mathrm{EA}}^2 < q^2$) as two different behaviors are expected. In our case, q_{EA}^2 can be estimated to be around 0.4.

In the region $0 \leqslant q^2 \leqslant q_{\mathrm{EA}}^2$ the power law decrease (6.40) of the correlation is expected. To test this hypothesis, one defines for each L the quantities

$$\chi_L^{(s)}(q) = \sum_{x=1}^{L/2} x^s C_L(x|q), \tag{6.42}$$

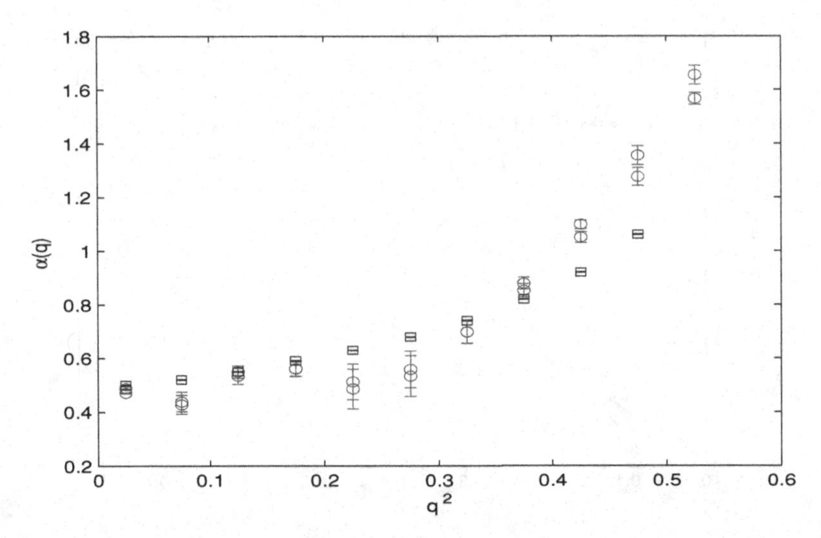

Figure 6.10 Circles represent the value of the exponent $\alpha(q)$ by fitting $\chi^{(s)}(q)_L$ as a power of L: for each value of q^2, two points are shown – corresponding to $s = 1$ and to $s = 2$. Squares represent the same quantity $\alpha(q)$, obtained through the scaling approach visible in Figure 6.11.

where $C_L(x|q)$ is the connected correlation function in a system of size L, i.e. $G_L(x|q) - q^2$. In order to decrease the statistical errors, one uses the asymptotically equivalent definition of the connected correlation function $G_L(x|q) = G_L(x|q) - G_L(L/2|q)$. For a large L, the quantity $\chi_L^{(s)}(q)$ should behave as $L^{s+1-\alpha(q)}$. The previous quantity for $s = 1, 2$ has been evaluated.

Instead of $C_L(x|q)$, its proxy $C_L(x|q) - C_L(L/2|q)$ has been used – it has smaller statistical errors. In the region $q^2 < 0.4$ it has been found that the ratio $\chi_L^{(2)}(q)/\chi_L^{(1)}(q)$ is well linear in L. Here the data for $\chi_L^{(s)}(q)$ can be well fitted as a power of L, and the exponents $\alpha(q)$ computed using $s = 1$ and $s = 2$ coincide within their errors. These results are no more true in the region $0.5 < q^2$, indicating that a power law decrease of the correlation is not valid there. The exponents found using this method are shown in Figure 6.10.

A different approach has been used to check these results for $\alpha(q)$. In the large-volume limit, the correlation function should satisfy the scaling

$$L^{\alpha(q)} C_L(x|q) = f(x/L). \tag{6.43}$$

The value of $\alpha(q)$ can be found by imposing this scaling.

The function $L^{\alpha(q)} C_L(x|q)$ has been plotted for each value of q, and the value of $\alpha(q)$ found for which the best collapse is obtained. The result of the collapse is shown in Figure 6.11 for q around zero where, for graphical purposes, it has been plotted as $L^{\alpha(q)} C_L(x|q) g(x/L)$ versus $\sin(\pi x/L)$, where the function g has been

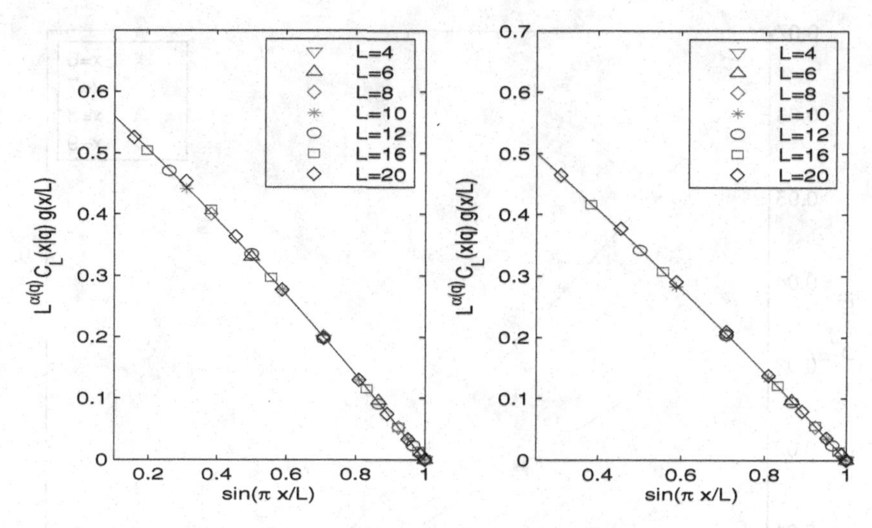

Figure 6.11 The quantity $L^{\alpha(q)}C_L(x|q)g(x/L)$ with $g(z) = (1/z + 1/(1-z))^{-\alpha(q)}$ versus $\sin(\pi x/L)$ for the set of data in the first bin $q^2 < 0.5$, using the best value of $\alpha(q)$. The left panel displays the data for all the correlation at distances $x \geqslant 1$ (the corresponding value of $\alpha(q)$ being 0.50); in the right panel, only the data with $x \geqslant 2$ are displayed (the corresponding value of $\alpha(q)$ being 0.54).

added to compress the vertical scale (it is convenient to use $g(z) = (1/z + 1/(1-z))^{-\alpha(q)}$). The left panel shows the collapse using all points with $x \geqslant 1$; in the right panel, the correlations at distance $x = 1$ are excluded. The corresponding values of the exponents are shown in Figure 6.10 and they agree with the ones coming from the previous analysis in the region of $q^2 \leqslant 0.4$.

The exponent $\alpha(q)$ is a smooth function of q^2 which goes to 1 near $q^2 = 0.4$, in very good agreement with theoretical expectations. There is no sign of a discontinuity at $q = 0$, and this is confirmed by an analysis with a high number of bins (e.g. 100). However, it is clear that for a lattice of this size one cannot expect to have a very high resolution of q, and one should look to much larger lattices to see if there are any signs of a discontinuity or plateau building up. The value of the exponent that was found at $q = 0$ is consistent with the value 0.4 found from the dynamics (Belletti *et al.* (2008)), and with the same value 0.4 found with the analysis of the ground states with different boundary conditions (Marinari and Parisi (2001)).

From the previous analysis, it is not clear if the exponent $\alpha(q)$ has a weak dependence on q, or if the weak dependence on q is just a pre-asymptotic effect. In order to clarify the situation it is better to look to the connected correlations themselves. Figure 6.12 displays the connected correlation $C_L(x|q)$ as a function of q^2 for $x = 6, 7, 8, 9$ at $L = 20$ (for the result at $x = 1$, see Section 6.3). The

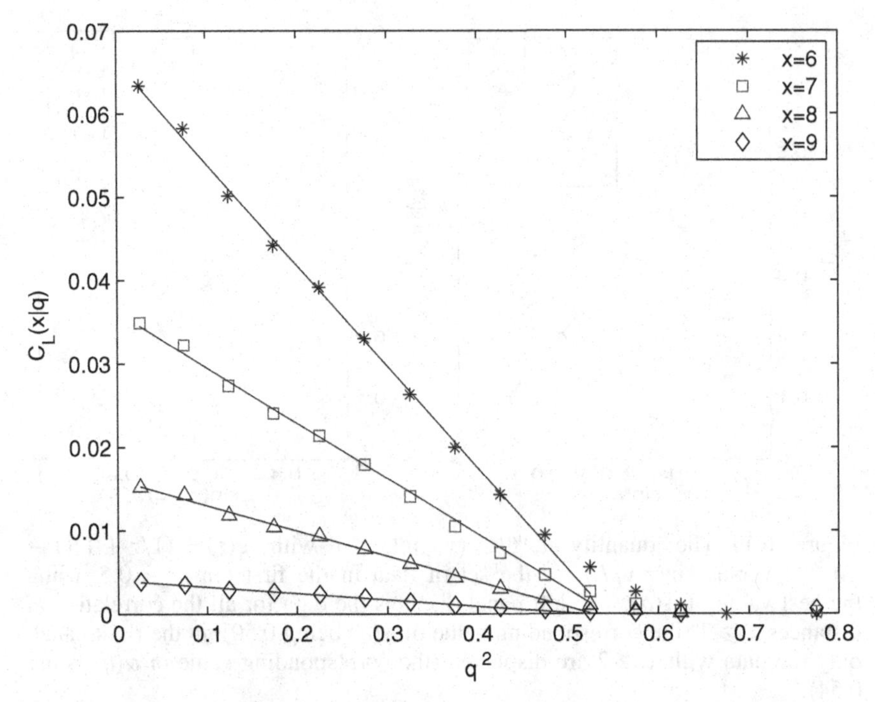

Figure 6.12 The connected correlation $C_L(x|q)$ as a function of q^2 for $x = 6, 7, 8, 9$ at $L = 20$. The straight lines are linear fits.

correlations at fixed L (e.g. $L = 20$) for large x has been fitted as:

$$C_L(x|q) = A(x, L)(q^2 - B(x, L)^2), \qquad q^2 < B(x, L)^2, \tag{6.44}$$

where $C_L(x|q)$ is very near to zero for $q^2 > B(x, L)^2$. The quality of these fits improves with the distance (similar results are valid at smaller L). The value of $B(x, L)^2$ is near to q_{EA}^2 and is slightly decreasing with L. The validity of the fits (6.44) for large L would imply that in the region $|q| < q_{\mathrm{EA}}$, the large distance decrease of the correlation function should be of the form $A(q^2 - q_{\mathrm{EA}}^2)f(x)$, and therefore the exponent $\alpha(q)$ should not depend on q.

However, near $q = q_{\mathrm{EA}}$ one should have a real crossover region. Figure 6.13 shows $C_L(x|q)g(x)$ at $L = 20$ for $0.475 \leqslant q^2 \leqslant 0.625$ versus $y \equiv (1/x + 1/(2L - x))^{-1}$ (the variable $y = x(1 - O(x/L))$ is used to take care of finite-volume effects), with $g(x) = (1 - 2x/L)^{-2}$. It seems that the data at $q^2 > 0.475$ decrease faster than a power at large distances and that the data at $q^2 = 0.475$ are compatible with a power law with exponent -1. It is difficult to extract precise quantitative conclusions without a careful analysis of the L dependence.

In the region $q_{\mathrm{EA}}^2 \leqslant q^2$ the task is different: the correlations are short-range and one would like to compute (if possible) the correlation length. In this region one

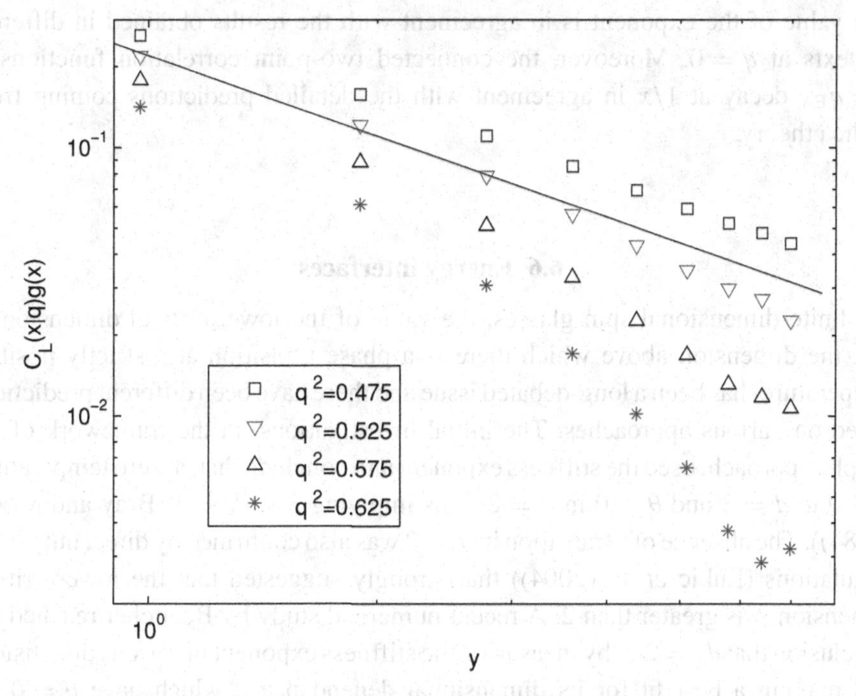

Figure 6.13 Correlations functions at $L = 20$ for $q^2 = 0.475, 0.525, 0.575, 0.625$, versus $y \equiv (1/x + 1/(2L - x))^{-1}$. The straight line is proportional to y^{-1}.

can fit the data as

$$C_L(x|q) = \frac{a}{x+1}\exp(-x/\xi_L(q)) + (x \to L - x) + \text{const.} \tag{6.45}$$

The choice of the fit is somewhat arbitrary, however it has only been used to check that the correlation length diverges at q_{EA}, and that near q_{EA} are well fitted by a $1/x$ power. The fits are good, but this may not imply correctness of the functional form in Eq. (6.45). It was found that far from $q = 0.5$ the correlation length is independent of L. A collapse of the data for $L > 8$ was tried in the form

$$\xi_L(q) = L\, f\big((q^2 - q_{\text{EA}}^2)L^{1/\nu}\big). \tag{6.46}$$

A reasonable collapse has been obtained, however the q_{EA}^2 is quite small (i.e. 0.25). It is quite possible that there are finite volume effects; also, different ways to evaluate q_{EA} give different results on a finite lattice, which would converge to the same value in the infinite-volume limit.

In conclusion, the global overlap for a two-clone system is a well-defined order parameter such that in the appropriate restricted ensemble, the two-point connected correlation function decays at large distances. The connected correlations decay as a power whose exponent seems to be independent from q for $0 \leqslant |q| < q_{\text{EA}}$.

The value of the exponent is in agreement with the results obtained in different contexts at $q = 0$. Moreover, the connected two-point correlation functions at $q = q_{\mathrm{EA}}$ decay at $1/x$ in agreement with the detailed predictions coming from replica theory.

6.6 Energy interfaces

For finite-dimensional spin glasses, the value of the lower critical dimension d_c (i.e. the dimension above which there is a phase transition at a strictly positive temperature) has been a long-debated issue and there have been different predictions based on various approaches. The initial investigations, in the framework of the droplet approach, used the stiffness exponent θ to conclude that, at zero temperature, $\theta > 0$ in $d = 3$ and $\theta < 0$ in $d = 2$, thus implying $2 < d_c < 3$ (Bray and Moore (1984)). The absence of a transition in $d = 2$ was also confirmed by direct numerical simulations (Lukic *et al.* (2004)) that strongly suggested that the lower critical dimension was greater than 2. A recent numerical study by Boettcher reached the conclusion that $d_c = 2.5$, by measuring the stiffness exponent in various dimensions and making a best fit for its dimensional dependence – which gave $\theta = 0$ for $d_c = 2.5$ (Boettcher (2005)).

An alternative approach has been put forward in Franz *et al.* (1994). To measure the stability of the low temperature spin glass phase against thermal fluctuations, Franz *et al.* suggested that the energy cost associated with imposing a different value of the overlap along one spatial direction be considered. Remarkably, using an analytical computation based on a mean-field approximation, they found the same result as when using the stiffness approach, i.e. $d_c = 2.5$. Such a prediction has been tested numerically on the Edwards–Anderson model in Contucci *et al.* (2011b). We will review here the numerical results of this work.

One considers two identical copies (same disorder) of a system constrained to have their mutual overlaps on the boundaries fixed to some preassigned values. More precisely, take two systems with the parallelepiped geometry of Figure 6.14. Each system has a base area L^{d-1} and height M; x denotes the coordinate along the longitudinal direction. Each of the two systems has free boundary conditions along this direction and periodic boundary conditions along the transverse direction. However, the two systems are coupled by the value of the overlap at their boundary planes. We compare the following two situations. In the first case, which we call the $(++)$ coupling condition, we impose an overlap $+1$ on both the left $(x = 1)$ and the right $(x = M)$ boundary plane; in the second case, which we will call the $(+-)$ coupling condition, the overlap on the left boundary $(x = 1)$ is equal to $+1$ and the overlap on the right boundary $(x = M)$ is equal to -1. We are interested

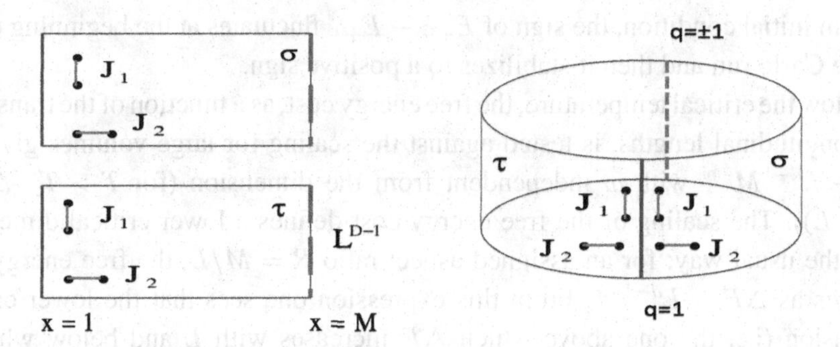

Figure 6.14 From two copies of longitudinal size M (lefthand side) to one system of longitudinal size $2(M-1)$ (righthand side).

in the difference of free energies between the two cases:

$$\Delta F = -\frac{1}{\beta}\mathbb{E}\left[\log Z_{+-} - \log Z_{++}\right] \tag{6.47}$$

where β is the inverse temperature and

$$Z_{ab} = \sum_{\sigma\tau} e^{-\beta(H_\sigma + H_\tau)}\delta\left(\frac{1}{L^{D-1}}\sum_{i\in A_1}\sigma_i\tau_i - a\right)\delta\left(\frac{1}{L^{D-1}}\sum_{i\in A_M}\sigma_i\tau_i - b\right) \tag{6.48}$$

with $a, b \in \{\pm\}$, A_1 the left boundary plane, A_M the right boundary plane, and H_σ the Hamiltonian of the spin configuration σ. One can prove that ΔF is a positive quantity thanks to the fact that $Z_{+,+} \geqslant Z_{+,-}$. Thus Eq. (6.47) seems an appropriate definition of a domain wall in spin glass models, in full analogy with usual ferromagnetic systems where a positive domain wall energy is obtained by imposing different values of the magnetization along one spatial direction.

Since free energies are not directly accessible through Monte Carlo methods, in our simulations we measured the differences between the $(+-)$ and $(++)$ cases of internal energies:

$$\Delta E = \mathbb{E}\left[E_{+-} - E_{++}\right] \tag{6.49}$$

where

$$E_{ab} = \sum_{\sigma\tau}(H_\sigma + H_\tau)\frac{e^{-\beta(H_\sigma + H_\tau)}\delta\left(\frac{\sum_{i\in A_1}\sigma_i\tau_i}{L^{D-1}} - a\right)\delta\left(\frac{\sum_{i\in A_M}\sigma_i\tau_i}{L^{D-1}} - b\right)}{Z_{ab}}. \tag{6.50}$$

We observe that at zero temperature $\Delta E = \Delta F$. Thus, at small enough temperatures, the positivity of ΔE follows from the positivity of ΔF by a continuity argument. In the numerical simulations that we have referred to, we always found a positive value of the energy differences at all temperatures. More precisely, for a random initial condition, the sign of $E_{+-} - E_{++}$ fluctuates at the beginning of the Monte Carlo run and then it stabilizes to a positive sign.

Below the critical temperature, the free energy cost, as a function of the transverse and longitudinal lengths, is tested against the scaling for large volumes given by $\Delta F \sim L^{d-1} M^{-\omega}$ with ω independent from the dimension (for $T > T_c$, $\Delta F \sim \exp(-L)$). The scaling of the free energy cost defines a lower critical dimension d_c in the usual way: for an assigned aspect ratio $R = M/L$, the free energy cost behaves as $\Delta F \sim L^{d-1-\omega}$. From this expression one sees that the lower critical dimension (i.e. the one above which ΔF increases with L and below which it vanishes with L) turns out to be $d_c = \omega + 1$. The mean-field computation of Franz *et al.* (1994) predicts $d_c = 5/2$, i.e. $\Delta F \sim L^{d-1} M^{-3/2}$. In other words, below the critical temperature, the free energy cost divided by the volume V turns out to be independent from L and, for large L, it scales as

$$\beta \Delta F / V \approx M^{-5/2} f(\beta). \tag{6.51}$$

Since the free energy difference and the internal energy difference scale with the same exponent below the critical temperature, we also have

$$\Delta E / V \approx M^{-5/2} f'(\beta). \tag{6.52}$$

By changing the boundary conditions at the phase transition point, one usually obtains

$$\Delta F \to C \neq 0.$$

Using Widom's scaling, we find

$$\frac{\Delta E}{V} \sim M^{-3+1/\nu}. \tag{6.53}$$

In a given case there could be cancellations and ΔF at the critical point could go to zero with the volume if $C = 0$. In Brézin *et al.* (2010) it was shown by an explicit computation that this does not happen, at least in the mean-field case (see Brézin (2010) for the ferromagnetic case). One expects that C is a smooth function of the dimension, so $C = 0$ in $d = 3$ would be surprising.

In the simulations, the above mentioned geometrical setup for the case of the Edwards–Anderson Hamiltonian with dichotomic (± 1) symmetrically distributed random couplings was considered. In particular, the implementation of the two systems coupled at their boundaries was obtained by considering a doubled-sized parallelepiped in the longitudinal direction. Denoting by y the coordinate in this direction of the doubled parallelepiped, the disorder variables are symmetric with respect to the central plane $y = M$. The subsystems in the half volumes $y < M$ and $y > M$ interact in the following way (again, see Figure 6.14):

Table 6.3 *System size parameters for the numerical simulations.*

L	M	Sweeps	Samples
4	4	8 192	9 600
	5	8 192	6 400
	7	8 192	6 400
	9	32 768	6 400
	13	262 144	3 200
	17	1 048 576	3 200
	21	4 194 304	3 200
	25	4 194 304	2 560

L	M	Sweeps	Samples
6	4	8 192	6 400
	5	8 192	9 600
	6	8 192	6 400
	7	8 192	6 400
	9	8 192	6 400
	13	32 768	3 200
	17	131 072	1 920
	21	1 048 576	1 920
	25	2 097 152	1 280

L	M	Sweeps	Samples
7	4	8 192	6 400
	5	8 192	6 400
	7	8 192	6 400
	9	8 192	6 400
	13	65 536	6 400
	17	524 288	3 200
	21	8 388 608	1 280
	25	8 388 608	1 280

L	M	Sweeps	Samples
8	5	8 192	6 400
	7	8 192	6 400
	9	8 192	6 400
	13	65 536	3 200
	17	524 288	1 920

- to keep the overlap of the spins σ and τ on the old $x = 1$ plane fixed to $+1$, the spins of the new doubled system are identified and they lie on the plane $y = M$;
- to keep the overlap of the spins σ and τ on the old $x = M$ plane fixed to either ± 1, the spins of the new doubled system at position $y = 1$ and $y = 2M - 1$ are identified. The strength of the couplings among the spins inside those planes are reduced by a factor of one-half (in order to have unbiased interactions); the couplings between nearest neighbors not lying inside those planes are flipped when going from the $(++)$ case to the $(+-)$;
- note that the effective longitudinal size of the total system is $2(M - 1)$.

The data described in Table 6.3 concern systems with sizes $4 \leqslant L \leqslant 10$ for the cubic lattice ($4 \leqslant L \leqslant 8$ and $4 \leqslant M \leqslant 25$ for the parallelepiped lattice) and with 30 different values of the aspect ratio $R = M/L$, ranging from 0.57 to 6.25. The computations have been performed by implementing a multi-spin coding of the parallel tempering algorithm, with 33 equally spaced temperatures ranging between 0.6 and 2.2.

Starting from the analysis of a cubic lattice ($M = L$), the data for the extensive internal energy difference were studied as a function of temperature for different

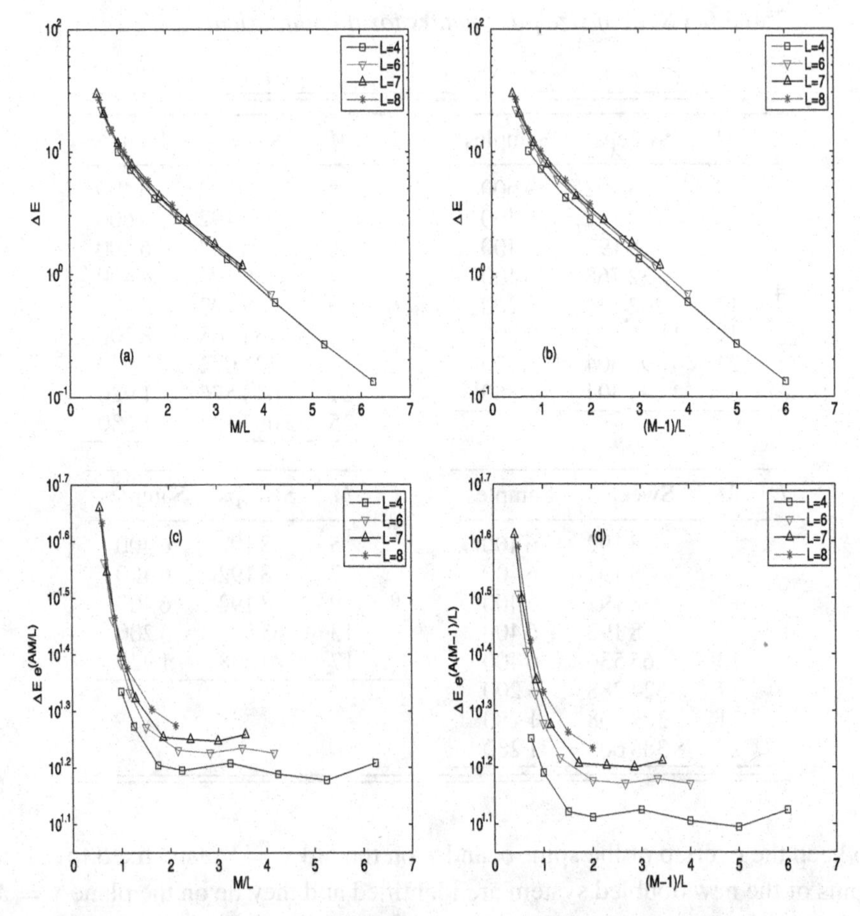

Figure 6.15 (a) ΔE as a function of M/L at temperature $T = 1.1$; (b) ΔE as a function of $(M-1)/L$ at temperature $T = 1.1$; (c) $\Delta E e^{AM/L}$ as a function of M/L; (d) $\Delta E e^{A(M-1)/L}$ as a function of $(M-1)/L$. A is 0.77. The different curves refer to different values of L.

system sizes. Below the critical temperature $T_c \approx 1.15$,x the data vary weakly with the linear lattice size L. The theoretical prediction would have been $\Delta F \sim L^{0.5}$ which suggests that there might be strong finite size effects on the energy interface cost for the small lattices we consider.

This is confirmed by the analysis around the critical temperature which is displayed in Figure 6.15. For the parallelepiped geometry, the extensive internal energy interface cost shows a different behavior when plotted against the aspect ratio $R = M/L$, or against the "corrected" aspect ratio $\tilde{R} = (M-1)/L$ (see Figure 6.15(a), (b)). Even though these two characterizations of the aspect ratios are totally equivalent in the thermodynamic limit, the data show a marked difference in the two cases. To appreciate this difference, (c) and (d) of the same figure show

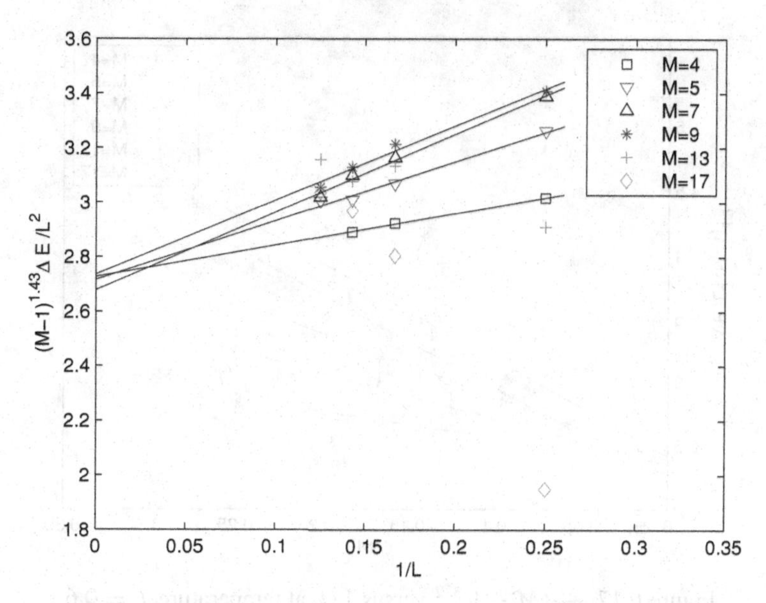

Figure 6.16 $\frac{\Delta E}{L^2}(M-1)^{1.43}$ versus $1/L$ at temperature $T = T_c$.

the same data multiplied by e^{AR} on the left and $e^{A\tilde{R}}$ on the right, where $A = 0.77$ has been computed by fitting the behavior of the data at large R (or $\tilde{R}$, it gives the same A) to an exponential form. An exponential dependence is not claimed here: the exponential re-scaling from the fit is just used to make the data variation more visible.

To extract sensible information from the data, a careful analysis of the dependence of ΔE from L and M is required. Figure 6.15 indicates that it is difficult to make an analysis of the data as a function of the aspect ratio (see also Carter *et al.* (2002)). Thus we start from the expected behavior of very large systems

$$\Delta E \sim L^2 f(M)$$

which is suggested by the general argument that each of the L^2 spins on each plane give the same contribution. Thus, for the small considered finite systems, the dependence on $1/L$ is studied for a fixed M. Since the purpose is to study the behavior of systems with a longitudinal size much smaller than their transverse size, the analysis must be limited to small values of M.

At the critical temperature, one expects the scaling of Eq. (6.53), i.e.

$$\frac{\Delta E}{L^2} \sim M^{-2+1/\nu}. \tag{6.54}$$

Figure 6.16 shows that $\frac{\Delta E}{L^2}(M-1)^{1.43}$ does not depend on L when L is large. This implies a value of $1/\nu = 0.57$, to be compared with $1/\nu = 0.408$ of Hasenbusch

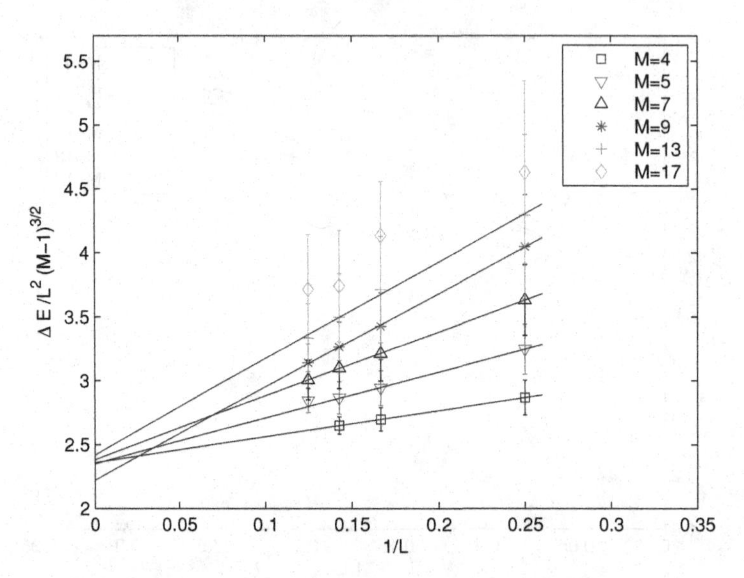

Figure 6.17 $\frac{\Delta E}{L^2}(M-1)^{3/2}$ versus $1/L$ at temperature $T = 0.6$.

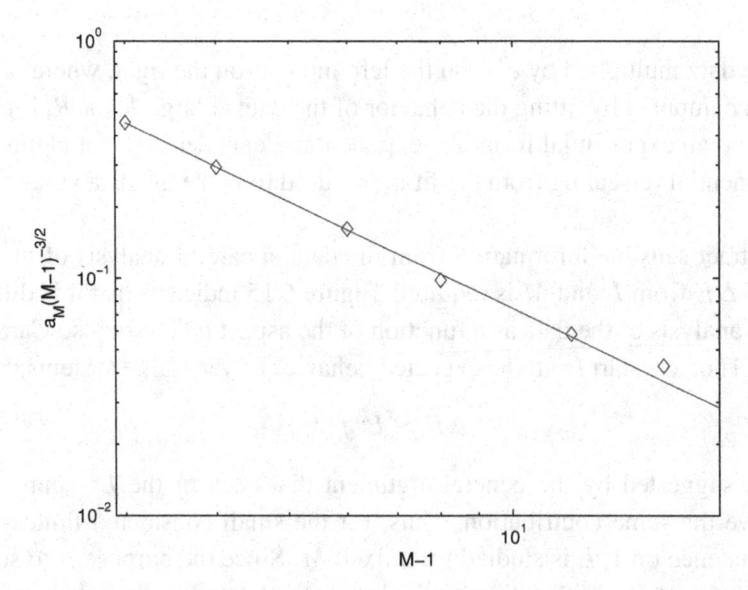

Figure 6.18 The intercepts of Fig. 6.17 as a function of $M - 1$ at temperature $T = 0.6$ in log–log scale. The straight line is the best fit of the form $a(M-1)^{-3/2}$.

et al. (2008). The predicted value of the exponent is 1.75; the difference between 2.45 and the predicted value (1.75) is not worrisome, given the fact that the value of ν was computed in Hasenbusch *et al.* (2008) using lattices (up to 40^3) that were much larger than those considered here and, historically, previous determinations of $1/\nu$ on smaller lattices gave large values of $1/\nu$.

Below the critical temperature, the scaling of Eq. (6.52) is expected, i.e.

$$\frac{\Delta E}{L^2} \sim M^{-3/2}. \tag{6.55}$$

In Figure 6.17 the quantity $\frac{\Delta E}{L^2}(M-1)^{3/2}$ is shown versus $1/L$. The behavior is described by a fit of the form $a_M + \frac{b_M}{L}$. The slope b_M increases with M while the intercept a_M is instead constant in M. In Figure 6.18, $a_M(M-1)^{-3/2}$ is shown versus $M-1$ in log–log scale. The data fall nicely on a straight line with a constant slope a. The best fit on b_M gives $k(M-1)^{-\delta}+h$ with $k<0$ and $\delta=-0.22$ if $M \in \{4,5,7,9,13\}$, while $\delta=-1/2$ excluding the values $M \geqslant 9$. Thus it would imply that the correction $\frac{b_M}{L}$ converges to zero for large volumes with an assigned aspect ratio.

Summarizing, in the regime $M \ll L$, the data for ΔE are well described by the following behavior

$$\Delta E \sim L^2(M-1)^{-3/2}\left(a+\frac{b_M}{L}\right) \tag{6.56}$$

which is compatible with the prediction $d_c = 2.5$.

References

Aizenman, M. and Contucci, P. 1998. On the stability of the quenched state in mean-field spin-glass models. *Journal of Statistical Physics*, **92**(5), 765–783.

Aizenman, M., Lebowitz, J.L., and Ruelle, D. 1987. Some rigorous results on the Sherrington–Kirkpatrick spin glass model. *Communications in Mathematical Physics*, **112**(1), 3–20.

Aizenman, M., Sims, R., and Starr, S.L. 2003. Extended variational principle for the Sherrington–Kirkpatrick spin-glass model. *Physical Review B*, **68**(21), 214403.

Aizenman, M., Sims, R., and Starr, S.L. 2007. Mean-field spin glass models from the cavity–ROSt perspective. Page 1 of: *Prospects in Mathematical Physics: Young Researchers Symposium of the 14th International Congress on Mathematical Physics, July 25–26, 2003, Lisbon, Portugal*. American Mathematical Society.

Almeida, J.R.L. and Thouless, D.J. 1978. Stability of the Sherrington–Kirkpatrick solution of a spin glass model. *Journal of Physics A: Mathematical and General*, **11**, 983.

Arguin, L.P. 2007. A dynamical characterization of Poisson–Dirichlet distributions. *Electronic Communication in Probability*, **12**, 283–290.

Arguin, L.P. 2008. Competing particle systems and the Ghirlanda–Guerra identities. *Electronic Journal of Probability*, **13**, 2101–2117.

Arguin, L.P. and Aizenman, M. 2009. On the structure of quasi-stationary competing particle systems. *Annals of Probability*, **37**(3), 1080–1113.

Arguin, L.P. and Chatterjee, S. 2010. Random overlap structures: properties and applications to spin glasses. *Arxiv Preprint arXiv:1011.1823*.

Barra, A. 2006. Irreducible free energy expansion and overlaps locking in mean field spin glasses. *Journal of Statistical Physics*, **123**(3), 601–614.

Bauke, H. and Mertens, S. 2004. Universality in the level statistics of disordered systems. *Physical Review E*, **70**(2), 25102.

Belletti, F., Cotallo, M., Cruz *et al.* 2008. Nonequilibrium spin-glass dynamics from picoseconds to a tenth of a second. *Physical Review Letters*, **101**(15), 157201.

Ben Arous, G., Dembo, A., and Guionnet, A. 2001. Aging of spherical spin glasses. *Probability Theory and Related Fields*, **120**(1), 1–67.

Ben Arous, G., Bovier, A., and Gayrard, V. 2002. Aging in the random energy model. *Physical Review Letters*, **88**(8), 87201.

Berestycki, N. 2009. Recent progress in coalescent theory. *Ensaios Matematicos*, **16**, 1–193.

Berretti, A. 1985. Some properties of random Ising models. *Journal of Statistical Physics*, **38**(3), 483–496.

Bhatt, R.N. and Young, A.P. 1988. Numerical studies of Ising spin glasses in two, three, and four dimensions. *Physical Review B*, **37**(10), 5606.

Bianchi, A., Contucci, P., and Knauf, A. 2004. Stochastically stable quenched measures. *Journal of Statistical Physics*, **117**(5), 831–844.

Billoire, A., Franz, S., and Marinari, E. 2003. On the tail of the overlap probability distribution in the Sherrington–Kirkpatrick model. *Journal of Physics A: Mathematical and General*, **36**, 15.

Boettcher, S. 2005. Stiffness of the Edwards–Anderson model in all dimensions. *Physical Review Letters*, **95**(19), 197205.

Bolthausen, E. and Kistler, N. 2009. On a nonhierarchical version of the generalized random energy model, II: Ultrametricity. *Stochastic Processes and their Applications*, **119**(7), 2357–2386.

Bolthausen, E. and Sznitman, A.S. 1998. On Ruelle's probability cascades and an abstract cavity method. *Communications in Mathematical Physics*, **197**(2), 247–276.

Bouchaud, J.P. 1992. Weak ergodicity breaking and aging in disordered systems. *Journal de Physique I*, **2**(9), 1705–1713.

Bouchaud, J.P. 2005. Spin glasses: the saga continues! *Journal Club for Condensed Matter Physics*. www.condmatjournalclub.org/wp-content/uploads/2007/06/jccm_novo5_02

Bouchaud, J.P. and Mézard, M. 1994. Self induced quenched disorder: a model for the glass transition. *Journal de Physique I (France)*, **4**, 1109.

Bovier, A. 2006. *Statistical mechanics of disordered systems: a mathematical perspective*. Cambridge University Press.

Bovier, A. and Kurkova, I. 2004a. Derrida's generalised random energy models 1: models with finitely many hierarchies. *Annales de l'Institut Henri Poincaré–Probabilités et Statistiques*, **40**, 439–480.

Bovier, A. and Kurkova, I. 2004b. Derrida's generalized random energy models 2: models with continuous hierarchies. *Annales de l'Institut Henri Poincaré–Probabilités et Statistiques*, **40**, 481–495.

Bovier, A. and Kurkova, I. 2006. Local energy statistics in disordered systems: a proof of the local REM conjecture. *Communications in Mathematical Physics*, **263**(2), 513–533.

Bovier, A., Eckhoff, M., Gayrard, V., and Klein, M. 2001. Metastability in stochastic dynamics of disordered mean-field models. *Probability Theory and Related Fields*, **119**(1), 99–161.

Bray, A.J. and Moore, M.A. 1984. Lower critical dimension of Ising spin glasses: a numerical study. *Journal of Physics C: Solid State Physics*, **17**, L463.

Brézin, E. 2010. *Introduction to Statistical Field Theory*. Cambridge University Press.

Brézin, E., Franz, S., and Parisi, G. 2010. Critical interface: twisting spin glasses at T_c. *Physical Review B*, **82**(14), 144427.

Cacciuto, A., Marinari, E., and Parisi, G. 1997. A numerical study of ultrametricity in finite-dimensional spin glasses. *Journal of Physics A: Mathematical and General*, **30**, L263.

Carter, A.C., Bray, A.J., and Moore, M.A. 2002. Aspect-ratio scaling and the stiffness exponent θ for Ising spin glasses. *Physical Review Letters*, **88**(7), 77201.

Chatterjee, S. 2009. The Ghirlanda–Guerra identities without averaging. *Arxiv Preprint arXiv:0911.4520*.

Contucci, P. 2002. Toward a classification of stochastically stable quenched measures. *Markov Processes and Related Fields*, **9**(2), 167–176.

Contucci, P. 2003. Replica equivalence in the Edwards–Anderson model. *Journal of Physics A: Mathematical and General*, **36**, 10961–10966.

Contucci, P. and Giardinà, C. 2005. Spin-glass stochastic stability: a rigorous proof. *Annales Henri Poincaré*, **6**(5), 915–923.

Contucci, P. and Giardinà, C. 2007. The Ghirlanda–Guerra identities. *Journal of Statistical Physics*, **126**(4), 917–931.

Contucci, P. and Graffi, S. 2004a. Monotonicity and thermodynamic limit for short range disordered models. *Journal of Statistical Physics*, **115**(1), 581–589.

Contucci, P. and Graffi, S. 2004b. On the surface pressure for the Edwards–Anderson model. *Communications in Mathematical Physics*, **248**(1), 207–216.

Contucci, P. and Lebowitz, J.L. 2007. Correlation inequalities for spin glasses. *Annales Henri Poincaré*, **8**(8), 1461–1467.

Contucci, P. and Lebowitz, J.L. 2010. Correlation inequalities for quantum spin systems with quenched centered disorder. *Journal of Mathematical Physics*, **51**, 023302.

Contucci, P. and Starr, S. 2009. Thermodynamic limit for spin glasses. Beyond the annealed bound. *Journal of Statistical Physics*, **135**(5), 1159–1166.

Contucci, P. and Unguendoli, F. 2008. Correlation inequalities for spin glass in one dimension. *Rendiconti Lincei, Matematica e Applicazioni*, **19**(2), 141–147.

Contucci, P., Graffi, S., and Isola, S. 2002. Mean field behaviour of spin systems with orthogonal interaction matrix. *Journal of Statistical Physics*, **106**(5), 895–914.

Contucci, P., Degli Esposti, M., Giardinà, C., and Graffi, S. 2003. Thermodynamical limit for correlated Gaussian random energy models. *Communications in Mathematical Physics*, **236**(1), 55–63.

Contucci, P., Giardinà, C., and Pulé, J. 2004. Thermodynamic limit for finite dimensional classical and quantum disordered systems. *Review in Mathematical Physics*, **16**(5), 629–637.

Contucci, P., Giardinà, C., Giberti, C., and Vernia, C. 2006. Overlap equivalence in the Edwards–Anderson model. *Physical Review Letters*, **96**(21), 217204.

Contucci, P., Giardinà, C., Giberti, C., Parisi, G., and Vernia, C. 2007. Ultrametricity in the Edwards–Anderson model. *Physical Review Letters*, **99**(5), 57206.

Contucci, P., Unguendoli, F., and Vernia, C. 2008. Lack of monotonicity in spin glass correlation functions. *Journal of Physics A: Mathematical and Theoretical*, **41**, 385001.

Contucci, P., Giardinà, C., and Giberti, C. 2009a. Interaction-flip identities in spin glasses. *Journal of Statistical Physics*, **135**(5), 1181–1203.

Contucci, P., Giardinà, C., and Nishimori, H. 2009b. Spin glass identities and the Nishimori line. Pages 103–121 of: *Spin Glasses: Statics and Dynamics Spin Glasses: Statics and Dynamics. Summer School Paris, 2007. A. Boutet de Monvel and A. Bovier (Eds.).* Birkhäuser, Basel-Boston-Berlin.

Contucci, P., Giardinà, C., Giberti, C., Parisi, G., and Vernia, C. 2009c. Structure of correlations in three dimensional spin glasses. *Physical Review Letters*, **103**(1), 17201.

Contucci, P., Dommers, S., Giardinà, C., and Starr, S. 2011a. Antiferromagnetic Potts model on the Erdos–Rényi random graph. *Arxiv Preprint arXiv:1106.4714*.

Contucci, P., Giardinà, C., Giberti, C., Parisi, G., and Vernia, C. 2011b. Interface energy in the Edwards–Anderson model. *Journal of Statistical Physics*, **142**, 1–10.

Contucci, P., Giardinà, C., and Giberti, C. 2011c. Stability of the spin glass phase under perturbations. *Europhysics Letters*, **96**, 17003.

Cugliandolo, L.F. and Kurchan, J. 1993. Analytical solution of the off-equilibrium dynamics of a long-range spin-glass model. *Physical Review Letters*, **71**(1), 173.

Cugliandolo, L.F. and Kurchan, J. 2008. The out-of-equilibrium dynamics of the Sherrington–Kirkpatrick model. *Journal of Physics A: Mathematical and Theoretical*, **41**, 324018.

Das, A. and Chakrabarti, B.K. 2005. *Quantum Annealing and Related Optimization Methods*. Vol. 679. Springer Verlag.

De Dominicis, C. and Giardinà, I. 2006. *Random Fields and Spin Glasses: a Field Theory Approach*. Cambridge University Press.

De Dominicis, C., Kondor, I., and Temesvari, T. 1998. Beyond the Sherrington–Kirkpatrick model. *Spin Glasses and Random Fields (Young, A.P. Editor), World Scientific*, 119–160.

De Dominicis, C., Giardinà, I., Marinari, E., Martin, O.C., and Zuliani, F. 2005. Spatial correlation functions in three-dimensional Ising spin glasses. *Physical Review B*, **72**(1), 014443.

Degli Esposti, M., Giardinà, C., Graffi, S., and Isola, S. 2001. Statistics of energy levels and zero temperature dynamics for deterministic spin models with glassy behaviour. *Journal of Statistical Physics*, **102**(5), 1285–1313.

Degli Esposti, M., Giardinà, C., and Graffi, S. 2003. Energy landscape statistics of the random orthogonal model. *Journal of Physics A: Mathematical and General*, **36**, 2983.

Derrida, B. 1980. Random-energy model: limit of a family of disordered models. *Physical Review Letters*, **45**(2), 79–82.

Derrida, B. 1981. Random-energy model: An exactly solvable model of disordered systems. *Physical Review B*, **24**(5), 2613–2626.

Derrida, B. 1985. A generalization of the random energy model which includes correlations between energies. *Journal de Physique Lettres*, **46**(9), 401–407.

Derrida, B. 1997. From random walks to spin glasses. *Physica D: Nonlinear Phenomena*, **107**(2-4), 186–198.

Derrida, B. and Gardner, E. 1986. Solution of the generalised random energy model. *Journal of Physics C: Solid State Physics*, **19**, 2253–2274.

Diestel, R. 2000. *Graph Theory*. Springer.

Dorlas, T.C. and Wedagedera, J.R. 2001. Large deviations and the random energy model. *International Journal of Modern Physics B*, **15**(1), 1–16.

Dyson, F.J., Lieb, E.H., and Simon, B. 1978. Phase transitions in quantum spin systems with isotropic and nonisotropic interactions. *Journal of Statistical Physics*, **18**(4), 335–383.

Edwards, S.F. and Anderson, P.W. 1975. Theory of spin glasses. *Journal of Physics F: Metal Physics*, **5**, 965.

Eisele, T. 1983. On a third-order phase transition. *Communications in Mathematical Physics*, **90**(1), 125–159.

Ellis, R.S. and Newman, C.M. 1978. Limit theorems for sums of dependent random variables occurring in statistical mechanics. *Probability Theory and Related Fields*, **44**(2), 117–139.

Fisher, D.S. and Huse, D.A. 1986. Ordered phase of short-range Ising spin-glasses. *Physical Review Letters*, **56**(15), 1601–1604.

Fisher, D.S. and Huse, D.A. 1988. Equilibrium behavior of the spin-glass ordered phase. *Physical Review B*, **38**(1), 386–411.

Fisher, M.E. 1964. The free energy of a macroscopic system. *Archive for Rational Mechanics and Analysis*, **17**(5),377–410.

Fisher, M.E. and Caginalp, G. 1977. Wall and boundary free energies. *Communications in Mathematical Physics*, **56**(1), 11–56.

Fisher, M.E. and Lebowitz, J.L. 1970. Asymptotic free energy of a system with periodic boundary conditions. *Communications in Mathematical Physics*, **19**(4), 251–272.

Franz, S. and Ricci-Tersenghi, F. 2000. Ultrametricity in three-dimensional Edwards–Anderson spin glasses. *Physical Review E*, **61**(2), 1121.

Franz, S. and Toninelli, F.L. 2004. Kac limit for finite-range spin glasses. *Physical Review Letters*, **92**(3), 30602.

Franz, S., Parisi, G., and Virasoro, M.A. 1992. The replica method on and off equilibrium. *Journal de Physique I*, **2**(10), 1869–1880.

Franz, S., Parisi, G., and Virasoro, M.A. 1994. Interfaces and lower critical dimension in a spin glass model. *Journal de Physique I (France)*, **4**(11), 1657–1667.

Franz, S., Mézard, M., Parisi, G., and Peliti, L. 1998. Measuring equilibrium properties in aging systems. *Physical Review Letters*, **81**(9), 1758–1761.

Franz, S., Mezard, M., Parisi, G., and Peliti, L. 1999. The response of glassy systems to random perturbations: a bridge between equilibrium and off-equilibrium. *Journal of Statistical Physics*, **97**(3), 459–488.

Franz, S., Leone, M., and Toninelli, F.L. 2003. Replica bounds for diluted non-Poissonian spin systems. *Journal of Physics A: Mathematical and General*, **36**, 10967.

Gandolfi, A., Newman, C.M., and Stein, D.L. 1993. Exotic states in long-range spin glasses. *Communications in Mathematical Physics*, **157**(2), 371–387.

Ghirlanda, S. and Guerra, F. 1998. General properties of overlap probability distributions in disordered spin systems. Towards Parisi ultrametricity. *Journal of Physics A: Mathematical and General*, **31**, 9149.

Giardinà, C. and Starr, S. 2007. Variational bounds for the generalized random energy model. *Journal of Statistical Physics*, **127**(1), 1–20.

Ginibre, J. 1970. In: *Cargese Lectures in Physics*. Gordon and Breach, New York.

Glimm, J. and Jaffe, A. 1981. *Quantum Physics: A Functional Integral Point of View*. Springer Verlag.

Griffiths, R.B. 1967a. Correlations in Ising ferromagnets. I. *Journal of Mathematical Physics*, **8**, 478–483.

Griffiths, R.B. 1967b. Correlations in Ising ferromagnets. II. External magnetic fields. *Journal of Mathematical Physics*, **8**, 484–489.

Guerra, F. 1972. Uniqueness of the vacuum energy density and van Hove phenomenon in the infinite-volume limit for two-dimensional self-coupled Bose fields. *Physical Review Letters*, **28**(18), 1213–1215.

Guerra, F. 1995. The cavity method in the mean field spin glass model. Functional representations of thermodynamic variables. *Advances in Dynamical Systems and Quantum Physics, S. Albeverio et al., eds, World Scientific, Singapore*.

Guerra, F. 2003. Broken replica symmetry bounds in the mean field spin glass model. *Communications in Mathematical Physics*, **233**(1), 1–12.

Guerra, F. 2006. The replica symmetric region in the Sherrington–Kirkpatrick mean field spin glass model. The Almeida–Thouless line. *Arxiv Preprint cond-mat/0604674*.

Guerra, F. and Toninelli, F.L. 2002. The thermodynamic limit in mean field spin glass models. *Communications in Mathematical Physics*, **230**(1), 71–79.

Guerra, F. and Toninelli, F.L. 2003. The infinite volume limit in generalized mean field disordered models. *Markov Processes and Related Fields*, **9**, 195–207.

Guerra, F., Rosen, L., and Simon, B. 1976. Boundary conditions in the $P(\varphi)_2$ Euclidean field theory. *Annales Institute Henri Poincare*, **15**, 213–334.

Hasenbusch, M., Pelissetto, A., and Vicari, E. 2008. The critical behavior of 3D Ising spin glass models: universality and scaling corrections. *Journal of Statistical Mechanics: Theory and Experiment*, **2008**, L02001.

Hed, G., Young, A.P., and Domany, E. 2004. Lack of ultrametricity in the low-temperature phase of three-dimensional ising spin glasses. *Physical Review Letters*, **92**(15), 157201.

Hukushima, K. and Nemoto, K. 1996. Exchange Monte Carlo method and application to spin glass simulations. *Journal of the Physical Society of Japan*, **65**(6), 1604–1608.

Iniguez, D., Parisi, G., and Ruiz-Lorenzo, J.J. 1996. Simulation of three-dimensional Ising spin glass model using three replicas: study of Binder cumulants. *Journal of Physics A: Mathematical and General*, **29**, 4337.

Jana, N.K. and Rao, B.V. 2006. Generalized random energy model. *Journal of Statistical Physics*, **123**(5), 1033–1058.

Jimenez, S., Martin-Mayor, V., Parisi, G., and Tarancón, A. 2003. Ageing in spin-glasses in three, four and infinite dimensions. *Journal of Physics A: Mathematical and General*, **36**, 10755.

Kahane, J.P. 1986. Une inegalité du type de Slepian et Gordon sur les processus gaussiens. *Israel Journal of Mathematics*, **55**(1), 109–110.

Kelly, D.G. and Sherman, S. 1968. General Griffiths' inequalities on correlations in Ising ferromagnets. *Journal of Mathematical Physics*, **9**, 466–484.

Khanin, K.M. and Sinai, Y.G. 1979. Existence of free energy for models with long-range random Hamiltonians. *Journal of Statistical Physics*, **20**(6), 573–584.

Kingman, J.F.C. 1975. Random discrete distributions. *Journal of the Royal Statistical Society. Series B (Methodological)*, 1–22.

Kingman, J.F.C. 1982. The coalescent. *Stochastic Processes and their Applications*, **13**(3), 235–248.

Kirk, G.S., Raven, R.E., and Schoeld, M. 1995. *The Presocratic Philosophers*. Cambridge University Press.

Kirkpatrick, S., Gelatt, C.D., and Vecchi, M.P. 1983. Optimization by simulated annealing. *Science*, **220**(4598), 671.

Krzakala, F. and Martin, O.C. 2000. Spin and link overlaps in three-dimensional spin glasses. *Physical Review Letters*, **85**(14), 3013–3016.

Kubo, R., Toda, M., and Hashitsume, N. 1991. *Statistical Physics*. Springer, Berlin.

Landau, L.D. and Lifshitz, E.M. 1969. *Statistical Physics*. Elsevier Ltd, Oxford.

Ledrappier, F. 1977. Pressure and variational principle for random Ising model. *Communications in Mathematical Physics*, **56**(3), 297–302.

Lukic, J., Galluccio, A., Marinari, E., Martin, O.C., and Rinaldi, G. 2004. Critical thermodynamics of the two-dimensional ±J Ising spin glass. *Physical Review Letters*, **92**(11), 117202.

Marinari, E. and Parisi, G. 1992. Simulated tempering: a new Monte Carlo scheme. *Europhysics Letters*, **19**, 451.

Marinari, E. and Parisi, G. 2001. Effects of a bulk perturbation on the ground state of 3D Ising spin glasses. *Physical Review Letters*,**86**(17), 3887–3890.

Marinari, E., Parisi, G., and Ritort, F. 1994a. Replica field theory for deterministic models: I. Binary sequences with low autocorrelation. *Journal of Physics A: Mathematical and General*, **27**, 7615.

Marinari, E., Parisi, G. and Ritort, F. 1994b. Replica field theory for deterministic models. II. A non-random spin glass with glassy behaviour. *Journal of Physics A: Mathematical and General*, **27**, 7647.

Marinari, E., Parisi, G. and Ruiz-Lorenzo, J.J. 1997. Numerical simulations of spin glass systems. *Spin Glasses and Random Fields (Young, A.P. Editor), World Scientific*, 59–98.

Marinari, E., Parisi, G., Ricci-Tersenghi, F., Ruiz-Lorenzo, J.J., and Zuliani, F. 2000. Replica symmetry breaking in short-range spin glasses: theoretical foundations and numerical evidences. *Journal of Statistical Physics*, **98**(5), 973–1074.

Mattis, D.C. 1976. Solvable spin models with random interactions. *Physics Letters A*, **56**, 421–422.

Mézard, M. and Montanari, A. 2009. *Information, Physics, and Computation*. Oxford University Press, USA.

Mézard, M. and Parisi, G. 2001. The Bethe lattice spin glass revisited. *The European Physical Journal B-Condensed Matter and Complex Systems*, **20**(2), 217–233.

Mézard, M., Parisi, G., Sourlas, N., Toulouse, G., and Virasoro, M. 1984. Nature of the spin-glass phase. *Physical Review Letters*, **52**(13), 1156–1159.

Mézard, M., Parisi, G., and Virasoro, M.A. 1987. *Spin glass theory and beyond*. World Scientific Singapore.

Mézard, M., Parisi, G., and Zecchina, R. 2002. Analytic and algorithmic solution of random satisfiability problems. *Science*, **297**(5582), 812.

Morita, S. and Nishimori, H. 2008. Mathematical foundation of quantum annealing. *Journal of Mathematical Physics*, **49**, 125210.

Morita, S., Nishimori, H., and Contucci, P. 2004. Griffiths inequalities for the Gaussian spin glass. *Journal of Physics A: Mathematical and General*, **37**, L203.

Morita, S., Nishimori, H., and Contucci, P. 2005. Griffiths inequalities in the Nishimori Line. *Progress of Theoretical Physics – Supplement*, **157**, 73–76.

Neveu, J. 1992. A continuous-state branching process in relation with the GREM model of spin glass theory. *Rapport Interne No 267, Ecole Polytechnique*.

Newman, C.M. 1997. *Topics in disordered systems*. Birkhäuser.

Newman, C.M. and Stein, D.L. 1992. Multiple states and thermodynamic limits in short-ranged Ising spin-glass models. *Physical Review B*, **46**(2), 973.

Newman, C.M. and Stein, D.L. 1996. Non-mean-field behavior of realistic spin glasses. *Physical Review Letters*, **76**(3), 515–518.

Newman, C.M. and Stein, D.L. 1998. Thermodynamic chaos and the structure of short-range spin glasses. *Progress in Probability*, 243–288.

Newman, C.M. and Stein, D.L. 2001. Interfaces and the question of regional congruence in spin glasses. *Physical Review Letters*, **87**(7), 77201.

Newman, C.M. and Stein, D.L. 2002. The state(s) of replica symmetry breaking: Mean field theories vs. short-ranged spin glasses. *Journal of Statistical Physics*, **106**(1), 213–244.

Newman, C.M. and Stein, D.L. 2003a. Finite-dimensional spin glasses: states, excitations, and interfaces. *Annales Henri Poincaré*, **4**, 497–503.

Newman, C.M. and Stein, D.L. 2003b. Ordering and broken symmetry in short-ranged spin glasses. *Journal of Physics: Condensed Matter*, **15**, R1319.

Nishimori, H. 1981. Internal energy, specific heat and correlation function of the bond-random Ising model. *Progress of Theoretical Physics*, **66**(4), 1169–1181.

Nishimori, H. 2001. *Statistical physics of spin glasses and information processing: an introduction*. Oxford University Press, USA.

Ogielski, A.T. and Morgenstern, I. 1985. Critical behavior of three-dimensional Ising spin-glass model. *Physical Review Letters*, **54**(9), 928–931.

Olivieri, E. and Picco, P. 1984. On the existence of thermodynamics for the random energy model. *Communications in Mathematical Physics*, **96**(1), 125–144.

Onsager, L. 1936. Electric moments of molecules in liquids. *Journal of the American Chemical Society*, **58**(8), 1486–1493.

Palassini, M. and Young, A.P. 2000. Nature of the spin glass state. *Physical Review Letters*, **85**(14), 3017–3020.

Panchenko, D. 2010a. A connection between the Ghirlanda–Guerra identities and ultrametricity. *The Annals of Probability*, **38**(1), 327–347.

Panchenko, D. 2010b. The Ghirlanda–Guerra identities for mixed p-spin model. *Comptes Rendus Mathematique*, **348**(3-4), 189–192.

Panchenko, D. 2011a. The Parisi ultrametricity conjecture. *Arxiv Preprint arXiv:1112.1003*.

Panchenko, D. 2011b. A unified stability property in spin glasses. *Arxiv Preprint arXiv:1106.3954*.

Parisi, G. 1979a. Infinite number of order parameters for spin-glasses. *Physical Review Letters*, **43**(23), 1754–1756.

Parisi, G. 1979b. Toward a mean field theory for spin glasses. *Physics Letters A*, **73**(3), 203–205.

Parisi, G. 1980a. Magnetic properties of spin glasses in a new mean field theory. *Journal of Physics A: Mathematical and General*, **13**, 1887.

Parisi, G. 1980b. The order parameter for spin glasses: A function on the interval 0-1. *Journal of Physics A: Mathematical and General*, **13**, 1101.

Parisi, G. 1980c. A sequence of approximated solutions to the SK model for spin glasses. *Journal of Physics A: Mathematical and General*, **13**, L115.

Parisi, G. 1983. Order parameter for spin-glasses. *Physical Review Letters*, **50**(24), 1946–1948.

Parisi, G. 2001. Stochastic stability. Pages 73–80 of: *AIP Conference Proceedings*.

Parisi, G. 2004. On the probabilistic formulation of the replica approach to spin glasses. *International Journal of Modern Physics B*, **18**, 733–744.

Parisi, G. 2006. Spin glasses and fragile glasses: Statics, dynamics, and complexity. *Proceedings National Academy Sciences*, **103**(21), 7948–7955.

Parisi, G. and Ricci-Tersenghi, F. 2000. On the origin of ultrametricity. *Journal of Physics A: Mathematical and General*, **33**, 113.

Pastur, L.A. and Figotin, A.L. 1978. Theory of disordered spin systems. *Theoretical and Mathematical Physics*, **35**(2), 403–414.

Pastur, L.A. and Shcherbina, M.V. 1991. Absence of self-averaging of the order parameter in the Sherrington–Kirkpatrick model. *Journal of Statistical Physics*, **62**(1), 1–19.

Perman, M., Pitman, J., and Yor, M. 1992. Size-biased sampling of Poisson point processes and excursions. *Probability Theory and Related Fields*, **92**(1), 21–39.

Pitman, J. 2006. *Combinatorial Stochastic Processes.* Lectures Notes in Mathematics, Ecole d'Eté de probabilités de Saint-Flour XXXII-2002. Vol 1875. Springer, Berlin.

Pitman, J. and Yor, M. 1997. The two-parameter Poisson–Dirichlet distribution derived from a stable subordinator. *The Annals of Probability*, 855–900.

Ruelle, D. 1987. A mathematical reformulation of Derrida's REM and GREM. *Communications in Mathematical Physics*, **108**(2), 225–239.

Ruelle, D. 1999. *Statistical mechanics: Rigorous results*. World Scientific.

Ruzmaikina, A. and Aizenman, M. 2005. Characterization of invariant measures at the leading edge for competing particle systems. *Annals of Probability*, 82–113.

Sachdev, S. 2001. *Quantum Phase Transitions*. Cambridge University Press.

Sherrington, D. and Kirkpatrick, S. 1975. Solvable model of a spin-glass. *Physical Review Letters*, **35**(26), 1792–1796.

Simon, B. 1993. *The Statistical Mechanics of Lattice Gases*. University Presses of California, Columbia and Princeton, New Jersey.

Slepian, D. 1962. The one-sided barrier problem for Gaussian noise. *Bell System Technical Journal*, **41**(2), 463–501.

Sollich, P. and Barra, A. 2012. Notes on the polynomial identities in random overlap structures. *Arxiv Preprint arXiv:1201.3483*.

Sompolinsky, H. and Zippelius, A. 1982. Relaxational dynamics of the Edwards–Anderson model and the mean-field theory of spin-glasses. *Physical Review B*, **25**(11), 6860.

Sourlas, N. 2005. Soft annealing: a new approach to difficult computational problems. *Physical Review Letters*, **94**(7), 70601.

Southern, B.W. and Young, A.P. 1977. Real space rescaling study of spin glass behaviour in three dimensions. *Journal of Physics C: Solid State Physics*, **10**, 2179.

Talagrand, M. 2002. Gaussian averages, Bernoulli averages, and Gibbs' measures. *Random Structures and Algorithms*, **21**(3-4), 197–204.

Talagrand, M. 2006. The Parisi formula. *Annals of Mathematics*, **163**(1), 221–264.

Talagrand, M. 2010a. Construction of pure states in mean field models for spin glasses. *Probability Theory and Related Fields*, **148**, 601–643.

Talagrand, M. 2010b. *Spin glasses: a challenge for mathematicians: cavity and mean field models*. Springer Verlag.

Temesvari, T. 2007. Replica symmetric spin glass field theory. *Nuclear Physics B*, **772**(3), 340–370.

Temesvari, T. and De Dominicis, C. 2002. Replica field theory and renormalization group for the Ising spin glass in an external magnetic field. *Physical Review Letters*, **89**(9), 97204.

Thouless, D.J., Anderson, P.W., and Palmer, R.G. 1977. Solution of 'solvable model of a spin glass'. *Philosophical Magazine*, **35**, 593–601.

Toninelli, F.L. 2002. About the Almeida–Thouless transition line in the Sherrington–Kirkpatrick mean-field spin glass model. *Europhysics Letters*, **60**, 764.

Toulouse, G. 1977. Theory of the frustration effect in spin glasses: I. *Communication Physics*, **2**(4), 115–119.

Van der Hofstad, R. 2012. Random graphs and complex networks. *Lectures notes in preparation (2012). See http://www.win.tue.nl/rhofstad/NotesRGCN.pdf.*

Van Enter, A.C.D. 1990. Stiffness exponent, number of pure states, and Almeida-Thouless line in spin-glasses. *Journal of Statistical Physics*, **60**(1), 275–279.

Van Enter, A.C.D. and de Groote, E. 2011. An ultrametric state space with a dense discrete overlap distribution: Paperfolding sequences. *Journal of Statistical Physics*, **142**(2), 223–228.

Van Enter, A.C.D. and van Hemmen, J.L. 1983. The thermodynamic limit for long-range random systems. *Journal of Statistical Physics*, **32**(1), 141–152.

Van Enter, A.C.D., Hof, A., and Miekisz, J. 1992. Overlap distributions for deterministic systems with many pure states. *Journal of Physics A: Mathematical and General*, **25**, L1133.

Vuillermot, P.A. 1977. Thermodynamics of quenched random spin systems, and application to the problem of phase transitions in magnetic (spin) glasses. *Journal of Physics A: Mathematical and General*, **10**, 1319.

Zegarlinski, B. 1991. Interactions and pressure functionals for disordered lattice systems. *Communications in Mathematical Physics*, **139**(2), 305–339.

Index